W9-DAF-947

JOHN OPIE

VIRTUAL AMERICA

Sleepwalking through Paradise

University of Nebraska Press | Lincoln and London

Acknowledgments for the use of illustrations
appear on pp. 207–10, which constitute an
extension of the copyright page.

⊗

Library of Congress Cataloging-in-Publication Data
Opie, John, 1934–
Virtual America: sleepwalking through
paradise / John Opie.
     p. cm.
Includes bibliographical references and
index.
ISBN 978-0-8032-3571-7 (cloth: alk. paper)
1. United States—Civilization. 2. National
characteristics, American. 3. Technology—
Social aspects—United States. 4. Virtual
reality—Social aspects—United States.
5. Authenticity (Philosophy). 6. Human
ecology—United States. 7. Human geography—United States. 8. United States —Environmental conditions. I. Title.
E169.1.O65 2008   973—dc22
2008000752

Set in ITC New Baskerville.
Designed by A. Shahan.

Contents

List of Illustrations   vi

Introduction   ix

1. Welcome to VirtuaLand: *Old Dreamworlds and the Power of a New Modernity*   1

2. Antique America: *Searching for Authenticity*   39

3. Human Kodaks in the Future Perfect: *Virtual America Embodied in World's Fairs*   83

4. Sleepwalking in America: *A Brief History*   107

5. Finding Authenticity: *Inhabiting Place in America*   149

Illustration Acknowledgments   207

Notes   211

Bibliography   229

Index   243

Illustrations

Paintings, photographs, and digital images

A. *The Book as Landscape* by David Johnson (1983)   i

B. *Spacious Skies . . . Leo's Diner* by S. Harris (1981)   ii

1. *The Farmers Home—Harvest* by Frances F. Palmer (1864)   2

2. *Mountain of the Holy Cross* by Thomas Moran (1875)   3

3. *Snake River in the Tetons: Ansel's View* by John Opie (2006)   5

4. *3-D Hyperbolic Visualization of Internet Topologies* by Young Hyun (2000)   9

5. *Visualizing Web Space as a 3-D Cityscape* by Tim Bray (2000)   11

6. *SimCity* image from *Sims 2 Open for Business*   15

7. *View from Mount Holyoke, Northampton, Massachusetts, after a Thunderstorm—The Oxbow* by Thomas Cole (1836)   53

8. Detail from landform map of the United States by Erwin J. Raisz   55

9. Photo of Catskill Mountains railway station, Haines Corners, New York (1902)   57

10. *Busy Day at Soda Spring, Manitou, Colorado* by L. C. McClure
    (ca. 1900)   58

11. *Yosemite Valley from Glacier Point* by ([Karl] William Hahn 1874)   60

12. *Horseshoe Fall, Moonlight, No. 79* by G. E. Curtis (1870)   63

13. *Among the Sierra Nevada, California* by Albert Bierstadt (1868)   65

14. *Falls of the Kaaterskill* by Thomas Cole (1826)   67

15. Engraving of Yosemite Valley (1874)   68

16. National Park Service photograph of Yellowstone Inn and
    auto (1922)   71

17. Photos of Shiprock near the Four Corners region, and desert
    scene in southeastern Utah, by John Opie (1969)   74

18. *Vishnu's Temple* drawing by William Henry Holmes (1895)   77

19. Photograph of the Administration Building and Main Basin at
    night (1893)   84

20. *The Corliss Engine, in Machinery Hall* drawing (1876)   85

21. Stevenson cartoons referring to the 1893 World Columbian
    Exposition (1980) and the entrance to Disneyworld (1971)   89

22. *SimCity* image, from a *Sims* 2 birthday party   97

23. *New York and Brooklyn* by Elizabeth Parsons and Rosemary Atwater
    (1875)   109

24. *A Midnight Race on the Mississippi* by H. D. Manning (1860) and
    The "Lightning Express" Trains: Leaving the Junction by F. F. Palmer
    (1863)   116

25. Map showing the Toledo, Ann Arbor, and Grand Trunk Railway
    and its connections (1881)   119

26. *Superior Street, Cleveland, from Presbyterian Church* by Adolph
    Karst (1873) and *The "Town Pump"* by Survey Associates
    (1908)   120–21

27. *Daniel Boone Escorting Settlers through the Cumberland Gap*, by
    George Caleb Bingham (1851)   133

28. *Westward Ho (American Progress)* by John Gast (1872)   135

29. Map of Fair Oaks, California (ca. 2001)   159

30. *The Rocky Mountains. Emigrants Crossing the Plains* by Fanny F.
    Palmer (1868)   172

31. *Bristlecone Pines* photograph by Claude Fiddler (2005)   178

32. *Bristlecone Pines* watercolor by Valerie Cohen (1998)   182

Maps

1. U.S. Geological Survey Map of U.S. Counties (2005)   184

2. Hydrological Unit Boundaries (1998)   185

3. Major Land Resource Area Boundaries (1998)   186

4. *Ecoregions and Metropolitan Areas* by Robert G. Bailey (2001)   187

5. Greater Yellowstone Ecosystem (2005)   189

Introduction

You are opening a simple book about our connections to home place in a world increasingly dominated by placelessness. This is neither a philosophical work nor a conventional history. It is a sequence of observations that build on one another. Or, better yet, a string of connected essays intended to persuade the author, and willing readers, to come to a common conclusion: that "sense of place" is an essential human quality. Sense of place also describes an authentic American identity.

While sense of place means a deeply personal inhabitation of a specific physical location, I also argue that it is uniquely served by participation in a new phenomenon: virtual reality in cyberspace. Virtual reality is defined here as an immersive digital environment. The idea of virtual reality also, however, clarifies mental images through which we Americans have explained ourselves, our national geography, and our nation, past and present.[1]

This book was written in my living room, either penciled on a pad of yellow ruled paper or typed into my aging laptop computer. A great stone fireplace dominates my 1930s wood-paneled living room, over which towers a beamed cathedral ceiling. The room's bay window looks out into an oak forest. I change my seat to follow the southern and western sun around the room. This room is my preferred home place inside a log-sided house seated in the fabled Indiana Dunes, along the southern coast of Lake Michigan. Outside lies a broad shaded lawn that crackles underfoot with last year's acorns. I hope to replace the alien grass with dune grass after the grandchildren grow up. Other houses, mostly inhabited summers, nestle in the woods. The summer people leave to free the outdoors; a brilliant red maple just outside my window highlights the fall colors; and winter's stark naked trees follow, joined by a brooding lake, nighttime's dark shapes, and silence.

I am a dinosaur who still scribbles with pencil on paper and who expects that a genuine home place includes concrete location in real space and time. All this is changing. With a click of the mouse, a tap on the keyboard, the screen lights up, and something freshly magical happens before us. We call up huge and dynamic nonphysical environments that have their own substance. In this book we explore not only physical place but also the contemporary world's most enthralling alternative home place, that of a virtual reality in the landscapes of cyberspace. Virtual reality is a parallel world that is not something different from concrete space, but instead is a species of the same genus, space, widely accepted, although electronic rather than physical. At worst virtual reality is the thin veneer of a wired society that is satisfied with immersive games, endless blogs, and Web sites beyond counting. Virtual reality can also offer heightened creativity available nowhere else and entry into infinitely repeatable world-sized simulations far more entrancing than any mere myth or symbol, tale in a book, or scenic vista.

We can all be like Columbus at his best, when he inhabited his imagination about the bizarre possibility of traveling west to meet the east. Gaming (simulation) technology can allow us to visualize things that have never been seen in the known world, as Columbus visualized what stood beyond the Western Sea. In 2005 astrophysicist Andrew J. S. Hamilton of the University of Colorado brought astrophysics and gaming together to fashion a "Black Hole Flight Simulator."[2] It visualized Einstein's equations for a twenty-three-minute journey into the impossible: "the other side of infinity." Visitors to the simulator at the Denver Museum of Nature and Science fell into a black-hole abyss, but instead of darkness they encountered a chaotic churn of light energized by a maelstrom of particle collisions—a wormhole leading visitors from the black hole to another universe: an unexpected Columbus-like journey to a New World.[3] Hamilton calls relativity "the ultimate terra incognita"—a phrase familiar to historians of exploration—"a strange trip following Einstein's math wherever it may lead." What Einstein described in his theory of general relativity, says Hamilton, "are forces of space and time literally outside the real world we know, or can know."[4] Similarly, the Western Sea of Columbus's time was like a black hole. Maps of sixteenth-century exploration told of unknown regions: "Beyond this point there be dragons." Columbus's heated imagination conjured up a fresh world fabricated, like a simulation, from jumbled pieces of the known world and fables of unknown lands.

Hamilton says the connection between gaming technology and science is a new frontier, the "future chalkboard of science." In this book the chalkboard is the American experience, and the chalk is virtual reality. What simulation has done for science can open a revolution for history. In this book we especially look at virtual reality in cyberspace as a superior model allowing us to reexamine our mental worlds, the nature of authentic existence, and how well or badly the two have been connected.

The American experience is like an archeological site, filled, layer by layer, with the buried strata of the past. This book, using words and illustrations as shovels, trowels, dust brushes, and sieves, seeks to uncover the life-and nation-shaping artifacts from these layers. The buried layers of archaeology also provided Freud with his favorite image for the inner territory of our minds. And, said Freud, these layers, like our dreamscapes, may rise into consciousness at any time.[5] We have spent our history morphing amid multiple parallel worlds. Here we will seek to identify the connections (or lack of connections) among our individual selves, an American identity, and the geography "out there," whether First Nature (the natural world), Second Nature (the metropolitan infrastructure), or a new Third Nature (virtual reality in cyberspace). We reverse the equation: cyberspace becomes our default habitat, while the physical world becomes its alternative. Virtual reality in cyberspace can provide a rich means for reinterpreting the American experience. Which has become more satisfying, for example, the virtual myth of the American West or its reality? We often doubt that we can find the real West hidden behind its invented images. It is in our human nature that we never can abandon our interior worlds; our mental mapping, which creates comfortable and reliable personal geographies; our myths about America's West; our invention of what place is desirable, such as Yosemite, or undesirable, such as Newark. As Yi-Fu Tuan puts it, "All people undertake to change amorphous space into articulated geography."[6] We are fortunate when our expectations match external data, keeping us attached to the external world.

We will also explore how we Americans have historically dreamed about creating a better life in daily ordinary existence, the locus of place, and how we have shaped this dream. I divide this dreamed-of better life into three parts: the Engineered America of our built environment, the Consumer America of our passion for material well-being, and the Triumphal America of our conviction that we are the exceptional model for the rest of the world. These three dreamworlds have encouraged American geographical placelessness and thus indifference to our existence in real places during real time.

Finally, we will explore "sense of place" (Last Nature) as the environmental surround for our personal identity, now in the enlarged context of virtual reality. We will look at what I call "the Resettlement of America" by examining how to successfully inhabit the particularity of place. We will explore a subject that is becoming more and more difficult to discern: how to find a home—a specific place to which we feel we belong—in our multifold and multipath global world. Fortunately, as we shall see, today we also have the larger vision offered by the environmental sciences, notably the notions of ecosystem and bioregionalism. If we can identify authentic place, we can position the authenticity of self—a reification of place and self through interaction. Here is where this becomes a simple book about an awesomely difficult subject.

Americans experience a poverty of physical space and real place. Cultural geographer Yi-Fu Tuan contrasts Americans with the Eskimos in their environment above the Arctic Circle. The Eskimo environment, he says, is in our view bleak, readily becoming a "white-out" when a blizzard blends land, water, and sky into an undifferentiated sameness. In what to us is a poorly articulated environment, the Eskimos, to survive, have accentuated their perceptual and spatial skills to see a useable world.[7] Tuan goes on to describe the limited vision of the first Europeans—Magellan and Cook—to cross the Pacific Ocean. They were unaware of subtle Polynesian navigation techniques, attuned to the directions of waves and the motions of the boat. To the Polynesians the Pacific Ocean was "a network of seaways linking up numerous islands, not a fearsome expanse of unmarked water." This network was invisible to Europeans.[8]

The great opportunity of cyberspace is that it brings all our American layers together into one virtual place.[9] Virtual reality is not merely a reconstruction but something new. As we surf the Internet, inhabit multiple existences in game simulations, and repeatedly reinvent our identities in e-mail blogs, we need no longer envy the nine lives of a cat. After cycling through America's virtual worlds in this book, we will work, finally, to secure a connection to another, different dimension called "home place." This we will explore in our quest for Authentic America.

Anticipating the possibilities offered by virtual reality, Yi-Fu Tuan adds that mental mapping is "the means by which I escape from my animal state of being." Tuan defines the human species as "an animal that is congenitally indisposed to accept reality as it is." Escapes, he argues, are imaginative reconstructions of reality, indispensable human activities: "While flights from our animal selves can be silly, preposterous, even cruel and dangerous, they can also create new and more expressive ways for humans to flourish."[10] Our immersion in invented worlds has decisively shaped our consciousness about ourselves, our society, and the natural world around us. In fact, we cannot see ourselves, or nature, outside of such constructed images.

Sleepwalking through Paradise

As a child I was never a sleepwalker. But my sister was. She would roam the familiar territory of the house, navigate around furniture, walk down the stairs and into the kitchen, open locked rooms, even play the piano. Fortunately, her mental map of home was a good match for her real house. When she woke up, she remembered hardly anything or nothing at all. To her it had never happened.

Yet another kind of sleepwalking takes hold when I go on automatic pilot to drive from one familiar place to another, to go from home to the grocery store, or the hardware store, the wine shop, Borders and Starbucks, and home again. On stretches of our commonplace routes we blank out. We depend upon a simulation—a virtual reality—embedded in our heads by running the same errands many times. We inhabit the mental simulation more than the tedious reality. We no longer see the actual road, we intuitively manage the traffic, and we almost unconsciously avoid surprises and accidents.

Our automatic pilots go even further: we strip out the natural world to navigate by man-made landmarks—WalMarts, McDonalds, and nearly identical gas stations. Out of the car we can be preoccupied and still find our way among the chains of shops and food courts in shopping malls. Our effort is almost unconscious, since the space we move in is so routine and commonplace. The natural world beyond the highway and shopping mall has become background noise. Any effort to extract meaning from nature becomes a modest mental idling about the green berm or new shrubbery.

We cannot celebrate this sleepwalking routine. Our senses have been compromised and deadened. America's First Nature, the natural world, is disappearing from view. Today we instead inhabit Second Nature, the engineered world, or infrastructure, of our cities and their ganglia of highways, rail lines, and airline routes. When I taught in Newark, New Jersey, I had a colleague whose awareness of the outside world had become dulled because he had grown up in and always inhabited the industrial confusion of northern New Jersey. He closed out the dreary urban scene as a means of self-protection against ugliness and boredom. But when he shut down his sensory equipment, he also drained dry his sense that the outside world held any vigor, excitement, or purpose. Sleepwalking tunes out rewarding opportunities in the outside world by denying its worth in any form. The thousand interstate miles between my home near Chicago and the Rocky Mountains are usually reviled as tedious flat rangeland. Rural America has fallen off the radar of most Americans, who have been trained to find glory only in mental landscapes of great mountains, red rock canyons, or iconic historic sites.

What a better American I would be—connected to nature—if I stopped sleepwalking. Can I look at neighborhood details that don't normally strike me when I walk my dog around the block? The great environmental essayist Loren Eiseley tells of a stroll he took through some woods while his dog Beau raced around in excitement: "I looked on, interested and sympathetic, but aware that the big black animal lived in a smell prison as I, in my way, lived in a sight prison." What if I turned off my automatic pilot when driving and discovered the natural world hidden behind the billboards? Nature writer Annie Dillard, living at Tinker Creek, describes her repeated attempts to poke through her mental blinders to see a brilliant real world, true and reliable beyond the constructs of her consciousness.[11] Dillard called herself a pilgrim to Virginia's Tinker Creek, but she was really looking for the means to inhabit Tinker Creek instead of sleepwalking through it.

There is hope for recovery. Geographer Deborah Tall had to visit a foreign country for her awakening. The Irish island, she writes, "was a place that had to be constantly attended to—one couldn't muddle along ignoring it. I felt as if I'd awakened after years of sleepwalking." She adds, "I longed to escape indistinctness, to feel the world as unavoidably real, even if ferocious." And more: "I needed to learn the plot and poetry of this place, the outlines of time passing on it, in order that it not be merely scenery."[12] More than a century ago the naturalist John Muir worried that Americans had already become sleepwalkers through our national parks: his beloved Yosemite was only seen through the lens of recreation—people demanding entertainment—while re-creation—people refreshed by wild nature—had been lost.

Indiana essayist Scott Russell Sanders suggests that if we fine-tune our awareness of a specific geographical location—inhabiting it as a participant, trusting the connection between place and self—we will enjoy a high level of satisfaction, a sense of completeness, even if only for rare moments. Such a home place can be in the natural world or in the built environment. Sanders hopes that we will not remain sleepwalkers in America's paradise: "What is that attunement of self and the world if not an intimation of paradise? I have felt it often, not only in the presence of moving water or ghostly moths or nervous deer, but also in shimmering trees, in meadows of stars, in grasses swept by wind, in a chorus of crickets, and not only meetings with nonhuman nature, but also in passages of music and poetry, in the elegant finds of science, in the sharing of food and talk with people I love; I have felt it indoors and out, in company or in solitude."[13] Such experiences cannot exist in the nowhere of limbo, nor as philosophical abstractions, but must be joined intimately to a particular place at a particular moment. Landscape historian John R. Stilgoe would add that such an epiphany cannot be planned, structured, or anticipated. Indeed, the right moment is likely to come while we are wandering, taking the unprogrammed walk or the random trip: "Ordinary exploration begins in casual indirection, in the juiciest sort of indecision, in deliberate, then routine fits of absence of mind."[14]

Both Stilgoe and Sanders tell of paradise found in ordinary places. We don't need pilgrimages to the grandeur of the Tetons in the Rocky Mountains or to sacred historic sites like Jefferson's Monticello. Paradise can be found in ordinary, homespun, and definitely local landscapes. Indeed, paradise can found almost anywhere, because we invest a particular place with its characteristics; the challenge is to connect the image cocooned in our heads with the reality. Stilgoe points us to our immediate surroundings: "Ordinary landscape fits like an old shoe, comfortably, without conscious notice by its wearer." He adds: "Nothing profound, nothing earth-shattering, but everything fitting into a private worldview."[15] He doubts that any place is so ordinary, like Middle America, that it must remain perpetually invisible; we only need to look hard enough and inhabit a place in a personal fashion.

In our modern day such serendipity is suspect. It is too personal, frighteningly intimate, the connection unreal and untrustworthy. As we dissect every experience, we also question the likelihood of a genuine connection between the meanderings of our brainpans and any reality out there. This was Einstein's dilemma for his entire adult life. He was stunned when he realized that his path-breaking 1905 papers presented a catastrophic challenge to any human certainty of the underlying reality of material existence.[16] Similarly, Werner Heisenberg worried over the principal of indeterminacy. Thomas Kuhn further upset the applecart by identifying paradigms—the ground rules of scientific inquiry—as dependent upon fickle society, changeable over time.

Paradise may be what we make of it, yet it also needs to be "out there" to be genuine. I have been fortunate: my childhood was spent in parklike Riverside, Illinois; my teenage epiphanies came in red rock Bryce and Zion national parks; and my current habitation is centered in the benevolent Indiana Dunes of Lake Michigan. In contrast, my twenty-three years in Pittsburgh and my eleven years in Maplewood, New Jersey, never took hold. Some places, for some people, work better than others. They have a "voice" that we can hear, if we listen instead of sleepwalking.

Virtual America

1

Welcome to VirtuaLand
Old Dreamworlds and the Power
of a New Modernity

Parlor Games

In the nineteenth century the finest room in a small-town home
was its parlor. The parlor window in Ohio, Iowa, or Kansas
might face out on an iconic landscape: vegetable garden, white
picket fence, rural dirt road, lush field of tall corn, and horses
grazing in a distant pasture. Agricultural prosperity drove the
American Dream. The same parlor wall also often held another
image within a picture frame: a lithograph of craggy mountains,
deep woods, and pristine lakes. The print copied German immigrant painter Albert Bierstadt's imagined Rocky Mountains.
Or his romantic depictions of fabled Yosemite Valley. Americans
clung tightly to sublime wilderness as the foundation for a national identity. One of the most popular reproductions came
from Thomas Moran's *The Mountain of the Holy Cross*, painted
in 1875, which combined the romantic sublime and God's benevolent presence welcoming Americans to the pristine West.

Later in the nineteenth century the family parlor would feature the stereopticon as another window into the mythic Great
Outdoors. Parlor viewers matched up dual photographs to mar-

vel at a three-dimensional image of Niagara Falls or Colorado's Garden of the Gods. It was like magic, a visual experience not equaled until the holograph of the mid-twentieth century. When Americans did venture outside in the 1880s, they saw themselves no longer as explorers, soldiers, or migrants but now as tourists and vacationers. They comfortably framed the outdoors through the windows of a train's Pullman car or of the sitting room of a resort hotel deep in the Rocky Mountains. Yet another frame would appear by the 1920s, when a vacationing family could peer through their automobile's windshield when they pulled over at a designated "scenic overlook" or "inspiration point."

1. Most Americans lived on farms or in rural villages until the 1920s. Our earliest, durable image of an idealized America still features agricultural prosperity, based on physical labor and natural abundance. Jefferson and Crèvecoeur identified the yeoman farmer as the model American. Most important was the individual farmstead, compared to Europe's peasant village. The image here is not the rustic settler's cabin but the dreamworld of the mature, successful farmstead, yet with wilderness on the horizon. This popular Currier & Ives print, one of many depicting happy rural prosperity, suggests a viewpoint from either a neighbor's window or behind a passing horse, like the distant one on the road.

2. Americans fervently believed that God was on their side. They were the Christian nation of Manifest Destiny. For many the discovery of a snowy cross on a great peak in the untouched Rocky Mountain wilderness confirmed God's special presence in America. God and Nature were merged into this single image that captured the nation's imagination. Thomas Moran's large painting was based on William H. Jackson's equally famous 1873 photograph, taken during the Hayden survey. Reproductions of it, like sacred shrines, adorned parlor walls across the country. Today the cross's snowfield has largely disappeared.

By the middle of the twentieth century another window had taken over, not in the parlor but in the family room. The bright glow of the television set joined the aptly named "picture" window overlooking a standard suburban lawn and street. Television and suburbia merged in *Leave It to Beaver* and *Father Knows Best*. The entire outside world marched across the family's screen, though strongly filtered by producers and advertisers. Americans began to experience a visual saturation that both trivialized the experience and demanded a higher power of discrimination than most gave to entertainment.

In addition, the wall behind the television may also have held Ansel Adams's bold photograph of the Snake River winding its way before the Grand Tetons. Even as I type these words, I can look up to this iconic masterpiece hanging on the stone facing of my fireplace. Then and now, it doesn't matter that few of us know the actual site. If we were to pause there, we might still conclude that Ansel Adams was closer to the truth. Indeed, some thirty years after Adams I pulled over at that designated Tetons overlook. I parked my rented car to photograph the same site as Adams had, now with a tall aspen intruding upon the view. We have always constructed layered images of the American West that decisively shape the real thing.

Cyberspace and Virtual Reality: *The Purple Rose Syndrome*

By the end of the twentieth century Americans had added another window that swept away all earlier windows: cyberspace and its virtual realities.

The combined power of screen, keyboard, mouse, and joystick is magical. The computer indulges my fancies, morning, afternoon, and evening. We can become absorbed in virtual reality—a phenomenon far greater than a painting, a photograph, a movie in a theater, the DVD at home, or IMAX. For most Americans television as a window to the world has seen its day:

3. Ansel Adams offered mid twentieth-century Americans photographs that were as iconic in their presentation of a pristine America as landscape paintings had been in the nineteenth century. No other photographer's work was as widely displayed in poster-size images on middle-class suburban walls. One of Adams's most popular images was of the Snake River's classic S curve, winding in front of Wyoming's dramatic Grand Teton Range. Adams's photos were magnets drawing Americans to natural vistas. I took the photo above at approximately the same location (a "scenic" auto pullout), but perhaps forty years later, now with an aspen tree and large pines intruding on the scene and partially obscuring the bend in the river.

it is now a pale shadow compared to Googling, mailing lists and listservs, blogs and multiuser domains, ever-multiplying Web sites, virtual tourism and Mapquest, on-line race-car games, Las Vegas poker, or a John Madden football game that allows me to enact a virtual struggle between my heroic Pittsburgh Steelers and those nasty Cleveland Browns.

The dynamic landscapes of cyberspace offer immersion in simulations that blur fact and fiction. Game players, with guns flashing, speed through dark corridors in strange castles. The Internet's virtual malls, chat rooms inviting multiple identities, auction houses, and information cafeterias are now the most robust dreamworlds that ever existed. Avatars, or cyber-surrogates (you can have more than one persona), fabricate ideal body types, extraordinary intellectual skills, and striking personalities. This stripping away of real location allows individuals to conceal their real self, including age, gender, physical condition, and geographical location.[1]

In the 1960s the prescient science fiction writer William Gibson coined the word *cyberspace* in his novel *Neuromancer*, describing a landscape where hackers traveled in a computer-network world as if it were the real world that they could feel with all their senses. This fiction has now become fact. It is a slippery path. We inhabit parallel worlds that often merge into each other. Remember the heroine played by Mia Farrow in Woody Allen's film *The Purple Rose of Cairo*, who projected herself into her favorite movies. On the one hand, our flesh-and-blood selves remain immersed in our immediate physical world of home, workplace, shopping, entertainment, and travel, inhabited by loved ones, friends, fellow workers, acquaintances, and crowds of strangers. On the other hand, cyberspace today is a place allowing entry into an idealized realm. We can label this the Purple Rose Syndrome.

Each time we fire up the computer, we whisk ourselves into cyberspace to inhabit a seductive alternative existence. William Gibson remarked, "Everyone who works with computers seems to develop an intuitive faith that there's some kind of actual space behind the screen."[2] MIT computer professor Sherry Turkle calls this the "Disneyland Effect," where the dreamworld makes ordinary life and the messy real world so disappointing that one can hardly wait to return to the dreamworld.[3] Neither Alice in Wonderland nor Dorothy in Oz, neither Aeneas nor Dante, has anything over us. Hugo Lindgren adds that such electronic experiences are "a profound indicator of where the entire world is heading. Online, off-line; reality, fantasy—these distinctions will cease to matter as more and more of us pass our time in virtual environments."[4] We pass through a gateway to find a new kind of space that expands our imagination. Virtual realities, islands in cyberspace, become real space that we inhabit. We replace ordinary existence with cyberspace, to create a stand-alone world that is superior to the natural world. Virtual reality is intrinsically attractive because it offers a geography not bound by space, time, or circumstances.

Cyberspace is Third Nature. We inhabit cyberspace, which pushes aside physical location and physical connections. First Nature is the ecosystems of the New World, which we Americans have barely inhabited, much less understood, for the last three hundred or more years.[5] Second Nature is our man-made infrastructure of metropolises, factories, slums, suburbs, and highways—infrastructure that made the natural world disappear from view, dismissed as background noise. Cyberspace, however, displaces space and place and time as nothing has before it.[6] Indeed, it provides profound alternatives to First and Second Nature because it defines itself spatially with direct equivalencies of space, place, and time.[7] The tension stems from the fact that cyberspace is not natural, but a series of electronic constructions created by designers and users. On the other hand, First Nature, under the guise of wilderness, often appears to exist only as a human construct.

Even the function of home is changing. John Leland of the *New York Times* reports: "In a $10 billion [home entertainment] industry, the stakes are high. Like the television set before it, the game console is now colonizing American living rooms and the lives therein."[8] Our window into cyberspace makes the living room transparent. The protective walls drop away. Home place once provided the location for reading a book or magazine, indifferently watching television, playing family board games, or simple conversation. It once provided a place for privacy, solitude, and personal renewal to face another day. These old ways seem quaint. The living room or family room has become

Command Central for a seamless interface with blogs and e-mails, personalized music, digital video, and photo programs.[9] Netscape, Amazon, and eBay were launched in 1995, followed by Google in 1998; Apple's iPod in 2001; and social-networking sites MySpace, Friendster, and Flockr in 2003 and 2004. In 2004 128 million American adults were on-line daily; about 60 million had high-speed connections at home.[10] Googling, iTunes, Podcasting, and TiVo-editing link people on the basis of shared interests rather than physical proximity. "Family" is redefined. Philosophy professor Albert Borgmann identifies a "self-imposed ghetto of tastes." The future promises more boutique entertainment. Cyberspace "can put you in touch with lots of people, but they're all your kind of people."[11]

Let's not forget the physical needs of cyberspace. Over several decades in the late twentieth century, at the cost of many billions of dollars, a vast array of telecommunication and computing infrastructure was strung along telephone poles, buried underground, snaked across ocean floors, hidden inside building conduits, and launched unseen into orbit above us. To which we patch in our individual computers. Cyberspace, ethereal and virtual as it is, has its platform on physical networks in real geographical space.[12]

Nevertheless, its premier identity is electronic. The World Wide Web opened its doors in the 1990s, the creature of computer scientist Tim Berners-Lee at the CERN (European Organization for Nuclear Research) laboratory in Geneva, Switzerland. He started the Web as a network allowing easy access to documents. Similarly, the Internet's e-mail, devised by India's Sabeer Bhatia, could only have appeared in cyberspace. Above all else, Berners-Lee and Bhatia created a public commons, but this time a commons blessed by no limits.[13] The original notion of a commons goes back to medieval villages, where sheep were given free access to grazing. "Commons" eventually became a metaphor for the power of private "rights" to a lim-

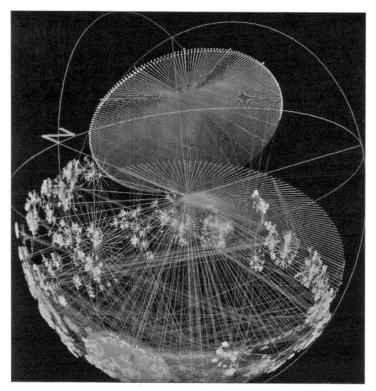

4. Both virtual reality and America's geography emphasize visual features. Walrus is a proprietary software tool allowing viewers to interactively visualize large phenomena in three-dimensional space. The developer, Young Hyun, uses three-dimensional geometry to display graphs through a fisheye-like distortion, its features deliberately shrunken or magnified depending on the point of view.

ited village site that soon was overgrazed. A virtual commons, in contrast, is infinite, and the old problems vanish. This virtual commons in which Americans are immersed is treated as the last, best place, marvelous in its reality and extraordinary in its potential. Some advocates nevertheless caution that real-world activities can bleed one-way into the virtual world.[14]

One is both in cyberspace and not in cyberspace. In the words of physicist Anton Zeilinger of the University of Vienna, because of modernity's confusion over the reliability of physical locality, "The world is not as real as we think." This leaves the advantage to cyberspace.[15] The real world, which must plod at its own pace, gives way to the heightened immediacy of virtual experiences. Virtual reality becomes the chosen world. If we die in a game, we are reborn, godlike, with one touch of a key. As a result we seem to enjoy a higher consciousness—free spirits liberated from ordinary space, time, and materiality.

When everyone is perpetually on-line, as I am with my cable modem, we can have immediate access to and constant contact with almost anyone, anywhere. We can enjoy instant gratification. We can live in an endless present that is not constricted by past or future. Everything is present-minded, happening *now*, ahistorical, and with scant resort to context. Unlike human affairs that twist their way through one-way historical time, "time's arrow," computer simulations are continuously repeatable and changeable, leaving no record. Cultural and historical context is missing from any given piece of information. As a participant I can enjoy, and must fear, a Promethean hubris.

We become gods, creating the universe anew. Mirroring today's cosmology, we live in an era of seemingly infinite expansion. Cyberspace is like the big bang at the beginning of our universe: it expands in a split second that unfolds, like origami, into multidimensional pure space.[16] Cyberspace provides a space of disembodiment and dislocation, where one's identity is defined by personal construction that makes traditional body codes (physical appearance, gender, age), community, and geographic place irrelevant. It is revolutionizing any remaining sense of belonging or place for Americans, with their vast mobility and indifference to a geographical location as "home."

Bruce Murray, former director of the Jet Propulsion Laboratory at Caltech, concludes that the Internet is "one of those transitions in human social structure that happens only once every few thousand years." Murray adds: "This period of time, this hundred or two hundred years of human history, is unprec-

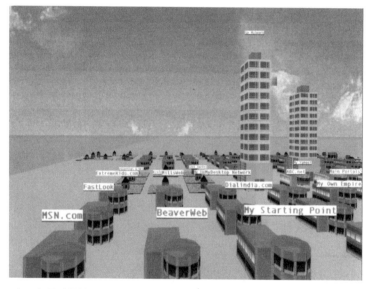

5A and 5B. A Web site is simultaneously its own reality and a virtual reality, thus challenging Americans to take a fresh look at their landscapes and their myths about these landscapes. The Web is both territory and map. This particular Web site has a double irony: it works only because it mimics the familiar urban neighborhood, with streets and shops, and the site itself, accessible as recently as late 2000, no longer exists in mid-2006.

edented. There's nothing like it, nor will there ever again be anything like it."[17] The Internet, as it whisks us through cyberspace, is seen as a revolution that takes us beyond the printing press, the harnessing of electricity, or the advent of television. Cyberspace advocates Martin Dodge and Rob Kitchin wrote in 2001, "Cyberspace is altering community relations and the bases for personal identity; is changing political and democratic structures; is instigating significant changes in organ and regional economies and patterns of employment; and is globalizing culture and information services."[18]

Simulacrum Games

Games are the bellwether for virtual reality. Whatever is happening happens there first—and to an extreme, quickly pushing the limits. Novelist William Gibson describes watching teenagers playing the early arcade games of the 1960s: "How rapt these kids were. . . . You had this feedback loop, with photons coming off the screen into the kids' eyes, the neurons moving through their bodies, electrons moving through the computer. And these kids clearly believed in the space these games projected." Will Wright, creator of *SimCity*, *The Sims*, and *Spore* (in the latter players create a new creature beginning with its biological origin), declares: "I think one thing that's unique about video games is not only that they can respond to you but down the road they'll be able to adapt themselves to you. They'll learn your desires. It might just be that games become deeply personal artifacts—more like dreams."[19] A human constant takes place here: everything on earth is a shadow of a primordial reality, now exquisitely played out on the computer screen.[20] An infinitely repeatable game is both a simulation and its own reality.

Those born between 1980 and 2000—the Millennials—are the chosen generation. They are especially intent upon games, either as solo players or with thousands of other identities involved in the same game on the Internet. Video games have grown into a huge business, soon expected to outpace the movie and television industries. These Millennials, in America alone, can now be numbered in the millions, spending tens of hours every week immersed in on-line games where they choose their roles or identities alongside many other similar players.[21] Long-standing Internet games such as *EverQuest*, with its Tolkeinesque medieval world, bring together several hundred thousand players creating virtual metropolises larger than Pittsburgh or Saint Louis or San Diego, and far larger than all of the cities of medieval Europe combined. Twenty percent of all on-line game players say they inhabit the game world as their "real" place; our real world is just where they eat and sleep. Some players, it is feared, could fall victim to what computer guru Edward Castronova calls "toxic immersion," in which their virtual lives take hold to the extent that their real-world lives collapse.[22]

Digital games, says editorial writer Hugo Lindgren, have "bulldozed childhood as we knew it."[23] Americans are more at home in these virtual realities than in any local geography circumscribed by a child's sandbox or tricycle, or an adolescent's basketball court or baseball field. The haunts of home are as likely to be found amid game scenery as in the neighborhood outdoors. The excessive multitasking nurtured in these games means that when children engage in schoolwork and other tasks, says psychologist David E. Meyer, "errors go way up and it takes far longer—often double the time or more—to get the jobs done than if they were done sequentially."[24] The article continues, "The brain needs rest and recovery time to consolidate thoughts and memories"—rest that becomes far less likely for "teenagers who fill every quiet moment with a phone call or some kind of e-stimulation."[25]

This bulldozing, however, is not necessarily as bad as it looks. Many parents, for example, are even supportive of *The Sims*, believing it superior to television and Barbie dolls (girls make up more than half of *Sims* users). More than sixty million copies have been sold, making *The Sims* one of the premier games. Players choose characters and create settings, both of which turn out to be largely identical to the conditions of their own suburban lives. As journalist Seth Schiesel observes: "As in real life, there are no points in The Sims and you can't 'win.' You just try to find happiness as best you can." One parent concludes: "The entire concept seems very creative. It seems as if it teaches them a lot about the different motivations and desires people have in life and shows some of the frustrations of running a household. In other games you see a lot of violence and we're not into that as a family. But it's interesting to see how they react to things with The Sims that normally a parent would have to deal with, like if one of their Sims doesn't want to go to school or is messy or if there are conflicting desires in the family." A twelve-year-old, describing why she switched from Barbie dolls, remarks: "You can't really develop the dolls. But in the Sims you're building the houses and putting the characters into different situations. . . . You can see what the characters are really doing. And also you can see how they get older and how they grow over time." Schiesel concludes: "The very characteristics that appear to make the Sims popular among girls—free-form game play and everyday setting—may also help explain why many girls stop playing the Sims once they actually start living young adult lives. The game as training ground gives way to the real thing."[26]

Boys, on the other hand, continue to use video games to refine their skills in technology. Writer Steven Johnson says that teenagers are "not using the technology to replace their real-world social life; they're using technology to augment it." He observes the "dramatic increase in cognitive engagement that the screen demands of them" and adds, "They're playing immensely complicated games, like *Civilization IV*—one of the

6. *SimCity*, which goes back to 1989, was designed by Will Wright, who found he was having more fun creating maps based on the game *Raid on Bungeling Bay* than he was playing the game itself. Instantly popular, *SimCity*, now in scores of versions and regularly updated, invented a new paradigm—a revolutionary new mode of perception—because unlike most games it could neither be won nor lost. Learning the intricacies of *SimCity* in all its versions is essentially a process of deconstructing how the software was organized to meet its specific set of goals and actions. The process was like "deep ecology," encouraging "deep simulation." The illustration, *Sims 2 Open for Business* (2006), allows immersion in one or more entrepreneurial ventures, from clothing boutiques to electronics shops to restaurant chains.

most popular computer games in the U.S. . . . in which players re-create the entire course of human economic and technological history." He concludes: "They're learning to analyze complex systems with many interacting variables, to master new interfaces, to find and validate information in vast data bases, to build and maintain extensive social networks crossing both virtual and real-world entitlements, to adapt existing technology to new ones. And they're learning all this in their spare time—for fun!"[27]

Yet all these analysts of the changing qualities of childhood still emphasize parental control; old-fashioned reading; physical play; and real-world contact with family, friends, and school. The games are still disembodied. Dinner with the family is more important. Human physical communication remains more powerful for bonding and meaning—facial expressions, body language, and other personal signals.

Computer games uncover *Homo ludens*, humanity at play. We turn to Johan Huizinga in his still-classic 1938 study. *Homo ludens* penetrates more deeply into our central being than does *Homo sapiens* or *Homo faber*.[28] Pure playfulness has an irreducible quality; it is indivisible, like pure place. Play expands our emotional cosmos: "The terrors of childhood, open-hearted gaiety, mystic fantasy and sacred awe are all inextricably entangled in this strange business."[29] Huizinga emphasizes the primordial quality of play. It is prior to language: most sports and computer games are nonverbal. It breaks away from normal activity to explore a deeper human consciousness. Indeed, it unveils "a second, poetic world alongside the world of nature."

Huizinga adds that play is "a stepping out of 'real' life into a temporary sphere of activity with a disposition all of its own."[30] It has its own rules and is self-validating, like a board game or field sport. The player loses consciousness of "ordinary reality." Some games are deliberately synthetic—fantasy worlds akin to *Lord of the Rings* or *Harry Potter*. Others are designed to be as lifelike and realistic as possible—*The Sims* or *Civilization* or games replicating World War II's D-Day. Both types involve battles and quests, with goals of virtual wealth and power.[31] Adolescents become tribal warriors who can lead their people through extremity and the unknown—descendants of Achilles, Odysseus, and Aeneas.

Huizinga emphasizes that a game-player, whether immersed in football or chess or playing on-line, becomes another being. The game is a magical experience for the participant—entry into the powerful world on the other side of the chasm. Whether in terms of Huizinga's play or modern games, we deliberately create new worlds that we believe enrich our lives through new forms of existence. These new forms are satisfying because they offer not only enjoyment but also new insights into the disturbing puzzle of our world.

One of Sherry Turkle's students at MIT told her about turning segments of his mind on and off, leading Turkle to note that "play and life are no longer a valid distinction."[32] In 2002 Edward Castronova identified an emerging overlap between synthetic and real worlds: real-world trading of in-game items flourishes in on-line marketplaces such as eBay. Virtual resources —whether swords, potions, or gold—are being sold for real money. Real businesses—IGE and UOtreasures—have been founded solely to buy and sell virtual goods.[33] People who are simultaneously playing several different virtual games transfer money from one world to another, as they would in a real-world currency exchange, or they buy game currency using real U.S. dollars or European Euros.[34] One *Project Entropia* player paid a real $26,500 in 2004 for an island in the game's virtual world. The Gross National Product (GNP) of the game *EverQuest*, when its "platinum pieces" appeared on eBay in 2001, was $2,266US per capita, richer than India or Bulgaria, making it the seventy-seventh richest country in the world—yet it didn't exist.[35] Those who respond by complaining that you can't eat virtual food to stay alive are reminded that most of a real diamond's value is virtual.[36] People treat their game property as though it's genuine personal property. If a game shut down, it conceivably would destroy hundreds of thousands of real marketable dollars. The Purple Rose Syndrome is a two-way street.

Simulations are little different from games. A Colorado River rafting trip down the Grand Canyon can be planned in detail using the *Grand Canyon River Trip Simulator* (GCRTsim), based on the diaries of 487 expeditions made along the 447-kilometer section of the river over an eighteen-month period. You learn how long it will take to have lunch before pausing for archaeological sites. Your rafting trip also depends on your type of raft, whether you use long or short oars, and whether you are motorized.[37] Simulated rafting through a virtual Grand Canyon provides immediately vivid rapids and instant access to flora and fauna. We're persuaded to trust the heightened simulations on the screen. Sherry Turkle is optimistic: "The voyager in virtuality can return to a real world better equipped to understand its artifacts."[38] She sees the computer as providing the best possible journey through the looking glass: the real experience might become tedious, only intermittently interesting, and largely disappointing. On the other hand, what cannot be simulated are real physical danger and real consequences. As my neighbor's bumper sticker says, "It ain't a sport if it can't kill you." So too with war games.

Most games celebrate a cowboy individualism, a strong libertarian bent that emphasizes a no-holds-barred entrepreneurship at best, and a bloodthirsty warrior mentality at worst. *EverQuest* claims to be egalitarian and purely free-market based, since everyone begins with few skills and no money. But those amassing huge fortunes quickly acquire the power to hold down the poor. The designers of the game *Ultima Online* morphed it into a modern welfare state that combines a free market with an activist government. Gaming today points to a gaming future based on a laissez-faire individualism with its own morality and mythology, brought on in part by the breakdown of traditional norms for behavior, as we shuttle between fantasy and reality.

Pushing the Envelope of the Known World: *Creativity and the Arts*

In VirtuaLand creativity and imagination seem enhanced as nowhere else. On-screen simulations offer opportunistic art forms, once the player gets past the stigma of the arcade game.

A Georgia Tech professor has likened "the multiple outcomes possible in video games to the magical realism of writers like Jorge Luis Borges."[39] Traditional media, such as literature and movies, are remade and reimagined. A *Chicago Tribune* reporter noted in November 2005, "Changing ways of accessing literature . . . could end up changing the way literature is produced."[40] Architect Marcos Novak, of the University of Texas at Austin, looks at cyberspace as "liquid architecture." It "breathes, pulses, leaps as one form and lands as another."[41] Liquid architecture creates an unprecedented new sense of place and space. Electronic painting, music, poetry, dance, fiction, and sculpture, including sex and sports, generate their own ecstasy or rapture—sublime epiphanies—that transports us beyond our ordinary life.

The history of art in all its forms is the continuous story of virtual realities. We can repeatedly and continuously construct and reconstruct visual and aural environments that mimic physical environments. Henry Jenkins of MIT writes: "What you need now is a garage band aesthetic, or independent film aesthetic for games. You're building the world from scratch. Why does it have to look like the world we live in."[42] Mimicing the real is not what interests most artists. Digital artists look for a world parallel to known reality, even for overlapping realities where borderlines disappear—rather than the familiar one.[43] Personalities in cyberspace exist simultaneously with their mundane selves in ordinary time and space.

Pioneers Dan Sandin, Tom DeFanti, and Caroline Cruz-Neira in 2005 created a prototypical large-scale virtual environment at the University of Illinois in Urbana. Using one computer tower, two laptops, two PowerPoint projectors, and a folding rear-projection screen, Sandin constructed an inhabitable work of art. Audiences shared virtual spaces in which they walk around, stand, and sit. These spaces were interactive. A visitor's look or stance or touch changed the virtual environment. In one exhibit, *Looking for Water*, participants began their adventure in outer space (based on real-time satellite images). They fell to earth to land on an archipelago where the lakes were fashioned from three-dimensional videos Sandin had made on a kayaking trip.[44] The participant was thus pulled from the known world into the displayed world, in this case a superphotorealistic world. Such virtual spaces morphed into the media-immersion pods that first appeared in Japan.[45]

Other artists, such as New York's John Simon, have devised software worlds that are not based on real-world images. Simon replaces realism, as he says, with "color theory, rhythmic motion, and automatic composition."[46] Mondrian is one of his models. Scott Rettberg, a media professor at Richard Stockton College of New Jersey, insists: "I see all these things as coexisting. It's not like one technology comes along and replaces another. . . .I think of electronic literature as a continuation of many different genres of experimental literature in the 20th Century. I never thought hypertext would eliminate the book. The book will never go away."[47] It's ironic that some of the best information about the Internet can be gleaned from old-fashioned newspaper stories.[48]

The Future Isn't What It Used to Be; or, Has Virtual Reality Always Been with Us?

We can profit from the realization that different iterations of virtual reality have always been with us. Over more than three hundred years we Americans have never been without complex mental configurations that we overlay like templates on the natural world. Then we work hard to force nature into those templates. Our challenge is to avoid spinning off wildly into fantastic realms. Sometimes we have failed, falling headlong into Hollywood, EPCOT, and Las Vegas. Our history, however, centers upon continuously shifting mental templates that require some connection with reality. They need to provide a kind of equivalence between signs and what they represent. As it is, confusion between myth and reality historically shaped western expansion across the Mississippi River. Immense harm was done when the Western Plains were oversold as a paradise, only to become the Dust Bowl. We now reconfigure the world with increasing ease, but we must also voice a healthy respect for "the world as it is." This latter obligation is one challenge addressed by the last chapter of this book.

Our history's virtual realities, held in the American *mentalité*, have long been far more vivid, enthralling, and real than the physical environments to which they supposedly refer. Carolyn Merchant recently reminded us that much of American history has been shaped by the attempt to recover the imagined Garden of Eden.[49] We revere our manufactured versions of the frontier, utopias, the West, natural parks, theme parks, and immigrant streets of gold.[50] The overwrought nineteenth-century landscape paintings of Albert Bierstadt, Thomas Cole, and Thomas Moran provided "mental mapping" that told Americans of near-sacred places that deserved pilgrimages. For many American travelers looking westward, their ambition was not struggle or adventure, but a personal encounter with an Edenic sublime. We choose to visit the aptly named Zion National Park in southwestern Utah because it has been defined for us as sublime. Yet, just outside the park entrance, the IMAX version of Zion National Park has become, for many visitors, more enthralling than the view from the canyon floor itself. Satisfied, they never enter the real park.

With virtual reality we witness, says Internet historian Margaret Wertheim, "the birth of a new [historical] domain, a new space that simply did not exist before." Wertheim calls virtual reality the digital version of Christianity's Heavenly City—Augustine's City of God—a place of pure knowledge untainted by materiality, "the realms of geometry and light, sparkling, insubstantial, laid out like a beautiful equation." Looking back in Western history to earlier versions of virtual reality, Wertheim explores Giotto's Arena Chapel in Padua as a medieval virtual world, in this case the walls covered with biblical scenes, "texts" in a day before widespread literacy: "We seem to be looking through the wall into a real physical space behind the picture plane. It is as if the archangel and the Virgin are really there in a little virtual world of their own beyond the chapel wall." Wertheim makes a direct connection: "Giotto created in the Arena Chapel a hyper-linked virtual reality, complete with an interweaving cast of characters, multiple story lines, and branching options." In this light Chartres Cathedral's stained-glass windows and Venice's Saint Mark's mosaics depict an autonomous "real world" that transcends miserable physical life. Hieronymus Bosch's visions of heaven and hell were perceived as windows into a reality world lurking behind superficial, ordinary life. For those few who could read, Dante's *Divine Comedy* "projected [one] into utterly absorbing alternative realities," notably an immaterial "soul space."[51]

Looking even further backward, the images on the cave walls at Altimira and Lascaux were windows into the magic of hunting.[52] Going to another extreme, in the Marquis de Sade's 1797 novel *Histoire de Juliette* the character Belmor muses with Juliette: "I do not know if reality measures up to the images we fashion from it. . . . Is not the pleasure I gain from this illusion preferable to that which reality brings me?" He continues, "It seems to me that I might do [ribald] things with . . . my imagination that the gods themselves would never admit." The French essayist André Breton may have been writing of the new wireless radio in 1924, but the shoe fits today's virtual reality. The new medium, Breton remarked, "gave me the illusion that I am embarked on some great adventure, that I somewhat resemble a seeker of gold: the gold I seek is in the air." In the 1930s the American impresario "Roxy Roth" bombarded the American public with the delights of Radio City Music Hall: "Two hours in the washed, ironized, ozoned, ultra-solarized air [of the Music Hall] are worth a month in the country."[53]

Americans in particular are prepared to inhabit cyberspace. Essayist Jonathan Koppell has put cyberspace in historical context, comparing it to the sixteenth century's unexpected and spectacular New World.[54] Cortez's entry into the New World made him a godlike apparition to Moctezuma: Aztec/Toltec mythology had prepared Moctezuma to see Cortez as a god. As for the Europeans, space itself had to be expanded and all the world's geography reinvented. Explorers and immigrants idealized the continent as either a new Garden of Eden, bringing a fresh start, or the final Paradise beyond the troubled Old World. In a similar vein virtual reality can be compared to Shangri-La, or Candide's El Dorado—a never-never land, a refuge from a desperate world, a paradise where nothing is wanting and no one ever grows old, and a magnet for the weary and discontented. We have a new generation of explorers, adventurers, pirates, entrepreneurs, immigrants, and heroes seeking their fortunes in cyberspace's New World. Cyberspace can be the new construction of America *ex nihilo*—the slate continuously wiped clean. Similarly, America, historically, has never been simply a new geography. It is another phenomenon that long existed alongside the known world. For centuries we explained America using the words and images of Europe. Cyberspace explains America afresh, including our classic mythologies about wilderness, frontier, and the West.[55]

Nineteenth-century visionaries who sought to inhabit the imagination included Ralph Waldo Emerson, Henry David Thoreau, and Walt Whitman. They read now as if they are our contemporaries. Emerson, for example, delivered up a fabulous image of the blessed immediacy of nature. No mere parlor windows for him: "Standing on the bare ground—my head bathed by the blithe air, and uplifted into infinite space,—all mean egotism vanishes." The self is merged into an alternative, better world: "I become a transparent eyeball. I am nothing. I see all. The currents of the Universal Being circulate through me: I am part or particle of God."[56] According to religious historian Cathy Albanese, "Nature might therefore have a quality of absoluteness about it," and thus, "Harmony with nature became the broad highway to virtuous living, and, more, to union with divinity."[57] This reads like the computer's push into cyberspace, the hurried rush into a simulation. Emerson might have been outraged or delighted.

Thoreau certainly delighted in the merger of the parallel worlds of self and nature—inner experience and outer reality: "Thus I caught two fishes as it were with one hook." He added, "It is very queer [fishing] in dark nights, when your thoughts had wandered to vast and cosmogonal themes in other spheres, to feel this faint jerk, which came to interrupt your dreams and link you to nature again."[58] Thoreau drops his line and hooks the formless void that stands behind the visible world. Literary historian Eric Wilson writes, "Thoreau's mind is like a plant, a supple, coherent form arising from the mire: he descends and ascends (rooting into dark earth, rising to the sun); draws nourishment from above and below (drinking from the stream bed and the stars); is passive, taking in carbon dioxide and light (no more busy with his hands and feet than necessary) and active, sending forth oxygen to alter the atmosphere (like a mining tool changes the earth)."[59] By drawing us into the extravagance of nature, Thoreau intended to prepare Americans for personal connection with authentic wilderness. Instead he trained us to enjoy the extravagance of cyberspace's Third Nature.

Virtual reality would also have found a home in Walt Whitman's personal cosmos. Whitman's *Specimen Days* talks of "the solid marrying the liquid" and "blending the real and ideal."[60] Whitman surfed Brooklyn, the East River, and Manhattan by means of ferryboats and horse-drawn buses. The pleasure was "in merely circulating." Like Emerson he abandoned "mean egotism" and opened himself utterly to the cosmos: "Mine is no callous shell / I have instant conductors all over me whether I pass or stop."[61] As in cyberspace "he would see parts and particles as circulations of a capricious whole. Anything can happen, and it usually did." Eric Wilson writes that Whitman's cosmos is "a boundless abyss in which atoms move in and out of temporary federations" like Web sites.[62] His "Song of Myself" affirms a larger-than-life Protean self that embraces the imagined dreamworld, also larger than life.

Emerson, Thoreau, and Whitman hold our attention because they were prophets, taking Americans into an unknown world where nothing can be taken for granted and where conventional responses fail us. They turned history into myth. In the same vein today's virtual reality throws new light on how we Americans have always filtered our experiences. There may be nothing wrong with cyberspace if we see it as discourse rather than any final reality.[63] Yet the inescapable synthetic abstractions of cyberspace can trivialize the physical reality of America's historical experience. It pushes the physical environment toward oblivion. Francis Bacon, the pioneering thirteenth-century scientist, recognized the problem: "God forbid that we should give out a dream of our own imagination for a pattern of the world."

Why Bother? *The Necessity of Mental Maps*

I once experienced the "Rosebush Exercise" with a psychologist. Imagine myself, I was told, in a favorite geographical place. Immediately, a Southwestern desert location came to mind. What was just in front of me? Small rocks on arid soil, small shrubbery, several juniper trees (my equivalent of a rosebush). In the middle distance? More junipers, rock outcroppings. On the horizon? Red rock canyon shapes that, like origami, unfolded forever. As well as an intensely blue sky. On both sides of me? Smooth large rocky shapes and more junipers. Behind me? A large rounded rocky formation. How large? About the size of a two-story farmhouse. "Aha!" said the psychologist. "That's very good." Apparently, I felt secure in my imagined place.

Of course I knew this all along. With luck we can have more than one place blessed with grace. I live year-round in a vacation community along the southeastern shore of Lake Michigan that became populated in the 1930s and 1940s by log-sided summer cabins built with enduring craftsmanship. They were judged throwaway cabins in the 1960s but by the 1980s had taken hold as treasured symbols of the village, which emerged as a concrete representation of a vacationist's escapist inner space. The cabins have since become scarce and thus even more adored.[64] The landscape is not only "out there." It can offer the concrete embodiment of a person's, and a society's, imagination.

Psychologists describe the virtual reality in our heads as *cognitive mapping*. We have an "inner space" that contains an internal geography. Our internal landscape is not helter-skelter, but our investment in an ordered mental image. Cognitive mapping, we are told, is vital to our personal well-being. We apply it, like a template, to the outside world. When our template cannot find a useable match with the external world, we become like Charles Dickens when he visited the United States and found the American wilderness outside his train window meaningless. He felt like he was aimlessly wandering in a random world. Mental mapping offers a "look-ahead" capacity that functions like over-the-horizon radar. It serves as an anchor point, a centering, necessary for individual self-identity. A virtual reality, say advocates, enhances and strengthens our internal map. Virtual reality in cyberspace plots out the "taste of a richer future."[65]

An elderly Frenchman, a former World War II Resistance fighter, Jacques Lusseyran, wrote that in his blindness, which had affected him since age eight, he enjoyed the "screen" his mind had constructed: "This screen was not like a blackboard, rectangular or square, which so quickly reaches the edge of its frame. My screen was always as big as I needed it to be. Because it was nowhere in space it was everywhere at the same time." So too with cyberspace. Lusseyran added: "Names, figures and objects in general did not appear on my screen without shape, nor just in black and white, but in all the colors of the rainbow. Nothing entered my mind without being bathed in a certain amount of light." Lusseyran's world was more than merely a window: "My personal world had turned into a painter's studio." When Lusseyran stood with a friend at a spot overlooking the Seine Valley, he compared his friend's sighted view with his "screen." He exclaimed that his friend's description held fewer pictures and not nearly as many colors: "When it comes to that, which one of us two is blind?"[66] In reporting this, the neuropsychologist Oliver Sacks said that Lusseyran felt that, in his blindness, he had "a deep attentiveness, a slow, almost prehensile attention, a sensuous, intimate being at one with the world which sight, with its quick, flicking facile quality, continually distracts us from."[67] A vote for Virtual America.

In another case detailed by Sacks a successful abstract artist whose mysterious disease had made him colorblind found himself "not only in an impoverished world, but in an alien, incoherent, and almost nightmarish one." He went through a slow and painful reconstruction of his visual world that turned a chaotic sensory flux into a stable perceptual world.[68] Another Sacks patient, an adult man who had long been blind, regained his sight following cataract surgery. He had been skillful and self-sufficient in a world that he touched with his hands, but a wooden cube from his touch world did not correspond to the cube he now saw. He could not exchange one reality for another, much less live in the parallel worlds of both realities. Or with the explosive impact of seeing colors: "[We] were now . . . demanding that he renounce all that came so easily to him."

Sacks writes, "The rest of us, born sighted, can scarcely imagine such confusion."[69] Like cube and color, cyberspace offers us a multitude of realities, each with its own cognitive domain, each with its own algorithms and rules.

Cognitive mapping grants us the world as we believe it to be, best when connected with reality, but many times with little correspondence. We are inquiring into the question of reliable human access to the physical facts of a landscape. We're disappointed by a vacation spot that doesn't fulfill our expectations, laid out by the colorful brochure. Too many folks take one disappointed look at the Grand Canyon from the El Tovar windows. One woman eating ice cream watched the sunset and complained, "Nothing so extraordinary about this, is there?"[70] She then continued on to Las Vegas.

The Quest for Authenticity: *Not All Virtual Realities Are the Same*

Neuropsychologist Oliver Sacks had his breakthrough when he asked what is "normal" and what is "not-normal" among the virtual realities that his patients inhabit. What is healthy well-being, and what is pathological? Sacks's deeply felt empathy for his patients' alternative worlds means he refuses to routinely damn their habitations.

Sacks describes a case that presents a disturbing parallel to today's preoccupation with computer simulations, with alternative selves on the Internet, in games, or in blogs. A severely brain-damaged young man was confined "to a single moment—the present—uninformed by any sense of a past or a future." Sacks describes this as "a profoundly pathological mental 'idling.'" The young man reacted immediately to all his experiences—"incontinently"—without perspective or judgment based on a larger worldview. He had no sense of "next," of anticipation, of intention, but rather was immersed in a "motionless, timeless, moment." To this patient "the grand, the trivial, the sublime, the ridiculous, are all mixed up and treated as equal."[71] He was the slave to every passing sensation. Sacks would find a time-free and placeless Virtual America a deeply pathological condition.

This alternative consciousness comes close to the childishness that the Frenchman Alexis de Tocqueville saw as deeply embedded in America's biography. This childishness is kin to American exceptionalism, the belief that the nation exhibited at birth its own full-fledged virtue and power. As he traveled through the United States in the 1830s, Tocqueville admired our simplicity in contrast to the tired worldliness of Europe. We were playful, cheerful, inventive, direct, and exuberant. But these are not Tocqueville's words; they are Sacks's, used to describe the young man whose brain idled in his single moment.[72] Tocqueville warned that the premier threat to American democracy was not tyranny, as in past despotisms, but infantilization, keeping Americans "fixed irrevocably in childhood."[73] Foreign visitors to America nevertheless admired the release of a playful impulse normally inhibited or lost in traditional cultures. Again, these are Sacks's words.[74]

Modern landscape historians John Brinkerhoff Jackson and John R. Stilgoe have both argued that a "landscape" means an invented space overlaid on geography—in Jackson's words, "a man-made system of spaces superimposed on the fact of the land." This overlaying is a habitual human activity. Jackson worries that we risk loss of true connection with true home place when we only see a world of stereotypes that conventionalizes our visual worlds—the same fast-food courts, the same motel rooms, even the same scenic view or inspiration point. Jackson calls for a primary distinction to separate the fact of the land from the illusory image evoked by a writer or artist, or tourist; or from the land transformed into a businessperson's resource, a politician's territory, a geologist's scientific data; or from the land displayed on a computer screen. In our contemporary world, he writes, we tend to inhabit "an existential landscape—without absolutes, without prototypes, devoted to change and mobility."[75] This existential landscape anticipated virtual reality in cyberspace.

Our entry into cyberspace—now a commonplace experience—involves the construction of a new American "normalcy"—a tantalizing Virtual America. Plato had Socrates at the end of "Phaedrus" tell of the king who wanted forgetfulness implanted in his citizens, to encourage people to rely on external influences rather than "the living speech graven in the soul."[76] There is no capacity, with no base-datum of normalcy, to distinguish between fantasy and reality. We can be stranded in the fantasy life of Virtual America. To use today's jargon, cyberspace becomes our "default habitat," the place where we start, which we prefer to inhabit, and by which we measure other habitats. Cyberspace acquires the character of genuineness. Freed from the messiness and shortcomings of the "real" world, paradoxically, virtual reality takes on a "realness" that makes it more vivid and inhabitable than the external world. If then the real world doesn't conform, we build walls of perception against that real world.

So what is right? Is there no American authenticity, but only a series of ephemeral virtual realities inhabited by idling sleepwalkers? In a 1998 book Allan Mazur describes the "Rashomon Effect" in the infamous case of deadly chemical pollution at Love Canal, outside Buffalo, New York. In the classic Japanese film *Rashomon* four individuals tell their vastly different stories about a murder along a forest path, and the viewer is left with the dilemma of who is telling the truth. Mazur uses the same approach to try to discriminate among the separate "realities" described by Hooker Chemical Company, by the local board of education that built a school on the site, by environmental activist and local housewife Lois Gibbs, and by local reporter Michael Brown. Mazur concludes that we cannot discriminate between what is factual and what is not. We are left bogged down in an "irresolvable morass of claims and counter claims."[77] Mazur wonders whether we can discover any final answer within such a multifold picture. Can we live without a guaranteed reliability—an authenticity—so evasive to our searching? Mazur faced a terrible problem in demanding absolute and clear finality. Without such finality no one was to blame for Love Canal. The "inordinate misery" caused by cancer, deaths, and suffering through exposure to chemicals at Love Canal would be neglected, or even forgotten.[78] The man had still been murdered on the forest path in Japan (yet the murder happened in a fictional movie).

Sacks has worked with the remarkable scientist Temple Grandin, who is autistic: "She perceived none of the usual rules and codes of human relationship. She lived, sometimes raged, inconceivably disorganized, in a world of unbridled chaos, destructive and violent."[79] What is normal, and what is impairment? I know someone who was probably autistic as a child. He struggled out of it with a lot of tender and knowledgeable care, invented a new self that he now inhabits with great success, and has no desire to return to what once was "natural" to him. Sacks notes that Grandin's pathology means she cannot understand ordinary social patterns. As an autistic person Grandin exhibits, says an amazed Sacks, a sort of personal intensity, clarity, and purity, "so far removed from the normal as to seem noble, ridiculous, or fearful to the rest of us."[80] She possesses "something deeply other." American exceptionalism fits the same mold.

Our mental map does more to shape how we perceive the outside world than the outside world does to shape our mental maps. Despite this imbalance we seek to connect ourselves to external prototypes—identifiable places and landscapes that form an authentic geography. According to the philosopher Mircea Eliade, this seeking reveals a profound need to be better rooted in a powerful landscape. We want to become safely centered in a concrete place and secured in physical reality.

Eliade calls this hierophany, or the existence of the sacred in our space and time. Theologian Rudolph Otto calls it the presence of "the holy," or the numinous "other."[81] We search for certainty in a world of flux and change.

Sometimes the sleepwalker awakens because of an epiphany. The external phenomenon takes over. The landscape historian John Conron argues that when we encounter an extraordinary sight like Niagara Falls, "we are awed, not informed, we settle for celebrating the sheer amazing fact that this wondrous thing is self-sufficiently there before us."[82] Social commentator Irvin Cobb wrote of the Grand Canyon: "You stand there, stricken dumb, your whole being dwarfed yet transfigured; and in the glory of that moment you can even forget the gabble of the lady tourist alongside of you who, after, searching her soul for the right words, comes right out and . . . pronounces it to be 'just perfectly lovely'!"[83] The essayist Charles Finger, when he entered Zion Canyon in southwestern Utah, felt "reduced almost to nothingness" by the dominant landforms soaring over him. Within the tight canyon walls, he wrote, "We were glad to come back to small, familiar things—the pebbled stream at our feet, and the moisture-loving ferns that grew in crannies near it. Those seemed to exist for our delight, but the tremendous sights [also] brought a sense of anxiety."[84] When Annie Dillard, in the 1970s, explored her mundane Tinker Creek in western Virginia, she broke through "the world of appearances" to enter Nature, a world of wonder, surprise, personal enlargement, and sacramental renewal.

The Darker Side of Virtual America

Not everything is happy in VirtuaLand. Virtual reality deserves to be celebrated, but also dreaded. It frees us from the uncertainties of daily existence by shifting our lives into a purer, happier world; it can also consume our lives through the loss of the external real world. There comes a loss of the richness and depth of the real thing—of its "aura" or "presence"—such as the real *Mona Lisa*; the real Wrigley Field; or the unbearable tension that accompanies the beginning of the real Olympic four hundred meters, compared to the delayed and edited broadcast.

For many Americans in 1999 the unsettling film *The Matrix* advanced simulation into a frightening and compelling virtual reality. It vividly portrayed ordinary daily life as a fiction programmed into the brains of millions of people sleeping in pods but fully believing they are experiencing normal lives. The "Matrix" penetrates their brains with a cable that inserts an interactive, simulated, but realistic world directly into their consciousness. All their thoughts and sensations flow from a computer code, a simulacrum of ordinary life. Behind all this are evil forces—bug-eyed machines—that guerrilla warriors undermine by unplugging promising human beings. Adam Gopnik of the *New Yorker* writes, "According to the rules of the movie, what is being destroyed is not real in the first place: the action has the safety of play [like a video game] and the excitement of the apocalyptic." He further observes that "few movies have had so much faith in their own mythology."[85]

Earlier stories of deception, in which "normal" human life is instead a virtual reality, include Kurt Vonnegut's novels *Slaughterhouse Five* and *The Sirens of Titan*. In the latter the entire history of humanity on the earth is simply a signal sent to restart a damaged spaceship marooned on Jupiter's moon Titan. Or consider Stanley Kubrick's *2001: A Space Odyssey*, in which HAL, the spacecraft's computer, doesn't deserve the reality he demands, and both a megalith on the moon and a space-coursing human fetus are intended to save humanity from itself. Similar questions are raised in other movies, including *Blade Runner*, *Total Recall*, and *Minority Report*.

Gopnik suggests a terrible confusion between reality and virtual reality. The meaning and reality of human existence are questioned:

> Although the movie [*The Matrix*] was made in 1999, its strength as a metaphor has only increased in the years since. The monopolization of information by vast corporations; the substitution of an agreed-on fiction, imposed from above, for anything that corresponds to our own reality; the sense that we have lost control not only of our fate but of our small sense of what's real—all these things can seem part of ordinary life now. . . . One can even start to wonder whether the language we hear on television and talk radio ("the war on terror," "homeland security," etc.) is a sort of vat [Matrix]-English—a language from which all earthly reference has been filed away. . . . We know what it's like to be captive to representations of the world that have, well, a faintly greenish cast. . . . [*The Matrix*] struck so deep not because it showed us a new world but because it reminded us of this one, and dramatized a simple, memorable choice between the plugged and the unplugged life. It reminded us that the idea of free lives is inseparable from the real thing.[86]

The boundaries, Sherry Turkle reminds us, are eroding between the real and the virtual. The border between the real and the virtual becomes permeable.[87]

As we inhabit the dark side of Third Nature, we also lose the unexpected, reject surprises, and deny involuntary experiences. The flaw of Virtual America is that it remains passive, mechanical, and impersonal, offering a bland, surface existence. It lacks any intrinsic imagination. Computer analyst Bruce Murray adds rather ominously: "The trends can't be sustained. . . . . We're in this remarkable transition—I call it the crunch. . . . . Outcomes are not a simple linear progression of what we see now. There are new alignments emerging—and eventually new belief systems. . . . In fact, we're at the limit. We could go crazy both individually and collectively, because of this rate of change. . . . There's a limit to how completely we'll be consumed by a cybersociety."[88] Would such a virtual reality run out of control, with no checks or any kind of governor?

Connections between internal self and the brilliance of the external world can void this Dark Side. Naturalists Annie Dillard, Aldo Leopold, and John Muir point to epiphanies that reshaped their imaginations and redirected their lives. In the moment of visionary rapture we enter an emotional geography. These are experiences without parallel. They are strange and anomalous. They cannot be programmed or anticipated. They can be kin to the seizures explored by Oliver Sacks. In them is a sense of revelation, of sacramental experience, a "doubling of consciousness," an unveiling of the "real" in the physical world. They bind together person and place. Sacks finds the recovery of the full quality of "pure" childhood experience—innocence, wonder, terror—in the surfacing of genius: Van Gogh, Poe, Kierkegaard, Lewis Carroll, Blake, Berlioz, Bunyan.[89] Dostoyevsky wrote, through Prince Miskin: "What if it is disease? What does it matter that it is an abnormal intensity, if the result, if the minute of sensation, remembered and analyzed afterwards in health, turns out to be the acme of harmony and beauty . . . completeness, of proportion?"[90] Today's virtual reality reminds us of the power of imagination over rough-and-ready ordinary life. Our "inner eye," electronically enhanced, is equally enlarged. The interactivity of the virtual window forges a personal intimacy with the external electronic world. Yet, as we shall see in the final chapter of this book, virtual reality cannot lead to authenticity: the irreducible uniqueness of the real individual person, inhabiting actual space and time, seeing through a glass clearly.

Keeping a Critical Stance inside Virtual Reality

My mechanical engineering students explore the functions of a worm gear in an on-screen simulation. The screen displays an idealized worm gear, compared to a greasy, gritty steel version that could be cradled in their hands. Similarly, medical simulations save lives by replicating torn or broken or diseased human bodies, and the procedures necessary to heal the wounded. Military simulations, senior cousins to fictitious battle games, go far beyond the war exercises that used to take place in the hot, dusty fields of Georgia, between red and blue armies. Edwin Link built a series of increasingly realistic flight simulations that trained cadres of military and civilian pilots. In all these cases valuable lessons are learned in a "safe" virtual environment. But—and the question applies to all of these—is it a mistake for the engineers never to have seen and handled and used a real worm gear?

We need to be continuously suspicious of cyberspace, applying a critical spirit established by modern Western science, based on independence and disinterestedness, truth and honesty, responsibility and duty. This requires more than ever a "sixth sense against the seductions of the spurious, the attractions of the ersatz."[91] The Disney Main Street versus the Indiana Main Street. An ethic can also emerge. Roger Kimball writes: "What responsibilities does a virtual world inspire? Virtual responsibilities, perhaps: responsibilities undertaken on spec, as it were. A virtual world is a world that can be created, manipulated, and dissolved at will. It is a world whose reverberations are subject to endless revision."[92] The delete key is always available. Whatever is done can be undone. Whatever is undone can be redone.

A profound contradiction plagues today's virtual realities. They are resoundingly effective in replacing "natural" encounters with the outside world, yet they present a huge opportunity, filled with unimaginable potential—to find in the virtual world a new way of living, of ordering one's world, even as the old ways are being swept away by uncertainty and placelessness. T. S. Eliot, in 1922, characterized the twentieth century as a "wasteland." Other writers today ascribe chronic triviality to the twenty-first century.[93] As Oliver Sacks points out, solitary internal experience alone becomes pathology.[94] Disneyland's Main Street becomes normative, yet it is unlike any original Main Street in any small town in Ohio or Kansas. How far can we go, without geography and history—First and Second Nature—to construct an authentic personal identity and a national identity? One possibility would be to embrace cyberspace as an Eden that transcends messy factuality. Emerson, in 1832, wrote, "Dreams and beasts are two keys by which we are to find out the secrets of our nature . . . they are our test objects."[95] The challenge is to hold together both dreamworlds and beastlike nature.[96]

Impermanence is written into cyberspace. Social critic Francis Fukuyama concludes that, because "anyone can get in and out of an Internet community whenever he or she wants . . . it doesn't lend itself to the development of shared norms and values, the sorts of things that really bind the more traditional kind of human community together." He questions the tensions that favor anonymity over accountability. "If you want to establish trust with other people, if you want to deal with them on a long-term basis, that's a social problem that isn't solved by technology. . . . If you want someone to trust you, you have to reveal something of who you are and establish your credibility. And that's not a problem that can be solved through technology."[97] Roger Kimball adds: "Culture survives and develops under the aegis of permanence. And yet instantaneity—the enemy of permanence—is one of the chief imperatives of our time." He writes, "We want what is faster, newer, less encumbered by the past."[98] The impermanence of virtual reality leaves no intrinsic meaning. Cyberspace doesn't point to anything that offers uniquely American permanence and reliability. Ephemeral passions—passing fads—have often overwhelmed the search for American qualities that can pass muster as normative and durable.

Cyberspace leads us to construct our society in new ways, just as landscape painting, photography, and television did in earlier days. Signs taken for reality may invidiously substitute for the real. "People explicitly turn to computers for experiences," says Sherry Turkle, "that they hope will change their ways of thinking or will affect their social and emotional lives"—through simulation games, fantasy worlds, or e-mails sent to a virtual community.[99] The political philosopher Hannah Arendt remarked that "what is at stake here [in the rush toward instant gratification] is the objective status of the cultural world, which . . . contains tangible things—books and paintings, statues, buildings, and music." Our judgment determines "their relative permanence and even eventual immortality," measured by "their most important and elemental quality, which is to grasp and move the reader or the spectator over the centuries." Americans lean toward passivity as media sponges, cheering hyped celebrities who are nothing more than hyped celebrities, watching heavily promoted television programs whose primary importance is that they have been heavily promoted. Americans especially are stuck inside the immediate fleeting moment in which, says Arendt, "there is no self, no value, no achievement, no criteria independent of the twitterings of fashion, which in turn are ultimately the twitterings of social power."[100]

For American culture to prosper, it must have authentic roots. It is one thing to have a psychological construction of the experiential world, to fabricate an independent world in hyperspace; it is another thing to have a healthy respect for the natural world as it is. And to interact properly with that natural world. Too often primary satisfaction resides in the mental construction. Virtual reality becomes objectified, the participant left mute or agnostic about the external world.

2

Antique America
Searching for Authenticity

I have a bad habit. I rarely go out of the house, bike a pathway, travel to a meeting, or journey to my favorite hideaway deep in Canyonlands, without a camera. I view my outer world through the 35-millimeter camera or its digital equivalent. My "normal" frame is that of a 100-millimeter lens, although I often zoom in and out for the best aesthetic effect. Is there an S curve in the image? Is the image divided in thirds? Is there depth in the image created by a nice framing foreground? Are there entrancing textures, colors, lighting, and shapes? I "see" in a vertical or horizontal frame, in a close-up or at a middle distance or toward a horizon, in panoramic view or in microdetail.[1] I force myself to ask whether I am seeing an image for its own sake; to satisfy my personal aesthetic sensibility; or, worst of all, to win a juried photo competition. In particular I like backlighting, reflections, vivid colors lit by the morning or afternoon sun, panoramics, and "natural" scenes with no human traces. Often the image in its final print looks better in monochromatic zones, from the blacks to the grays to the whites of Ansel Adams's famed zone system, instead of in the distinctive, even unreal colors of Kodachrome, Agfachrome, or Fujichrome.

If I am lucky, I can reshape my framed scene into an iconic abstraction with the help of computer software. I admit that I search for so-called sublime or picturesque landscapes, but there is a lot to be said for an ordinary city street. From the right viewing angle little is so "ugly" that it cannot be transformed into a startling image. My challenge is to break through yet another veil—the camera's viewfinder—that I've introduced between my personal sense and outside landscapes. Even when I don't have my camera with me, I still "frame" my visual world as if I were ready to take a picture.

Other veils prevail. My internal screen is captive to preconceptions, mostly American but also derived from Europe, with a touch of the Native American and Japanese. These shape the images chosen from my camera's window on the world, which is both window and veil. In my case I seek out images that, consciously and unconsciously, reflect the curving streets, substantial houses, and greenery of my hometown of Riverside, Illinois, or a sixteen-year-old's enthralled first impressions of Utah red rock country blessed by light, or the changing seasons along the shifting shoreline sands of the Indiana Dunes along Lake Michigan. In turn all these images are filtered by Cole and Bierstadt, Georgia O'Keeffe and Ansel Adams, as well as by the words of Crèvecoeur, Thoreau, Muir, Loren Eiseley, May T. Watts, Abbey, Dillard, and colleagues like Peirce Lewis and John Stilgoe. And a myriad of others. They taught me to seek out the sublime rather than the picturesque. They turned the ordinary into the sensational (in the latter word's original meaning). They taught me the difficult task, at least for me, of shifting from observer to inhabiter. Looking out the window invited me to the door.

Because I've learned to see the world through a camera's lens, I've chosen journeys to locations I might never otherwise have visited, met people I might never otherwise have met, and preserved memories of these experiences. My memories are wired into my photos—and recovered primarily through them. This enterprise is risky because it can be so limiting. But as George Shaub, the editor of *Shutterbug*, writes: "What I really like about travel photography is how it skins my eyes. It's what practitioners of Zen might call the 'slap of consciousness.'" He adds that his daily routine engenders sleepwalking, "a getting through the day mentality that stifles awareness and shuts the creative eye." The challenge is to break out of this stupor. Shaub remarks, "I travel with eyes wide open, I get fascinated by things like old doors and walls, how train tracks curve through old markets." He once took a photo class where the instructor's first assignment was for students to take photographs no more than fifty feet from their front door. The outcome was the transformation of mundane ordinary space into vibrant place: "She was trying to tell us that the world is filled with wonder, miracles that we fail to see because of the routines we follow." Shaub also connects this with travel: "Getting back on the road, even close by home, freshens my vision once again, with the camera being the vehicle for the visual ride."[2] The trick is to use the vehicle as a transparency instead of as a brick wall.

I light out for territory with the belief that it will insert me into an authentic place. I usually don't succeed because of my blinders. In another context, years ago, before I learned about environmental science, I started my Western Civilization freshman survey course by announcing, "Man no longer has biology, only history." This notion has since been overturned, and I have begun to teach differently. I'm still learning to think and live as an inhabitant of natural and built environments measured by the dynamics of local ecosystems, both badly broken and relatively healthy. If I am lucky, I can feel that I personally inhabit a place, that I find it extremely satisfying. Nevertheless, it is all too easy for me to slip into sleepwalking through paradise. I want a sense of immediacy, of intensity, of surprise in whatever place I inhabit. When this works, I am home. It is authentic, and I begin to find my identity.

Today, like many Americans, I also "see" with the help of Web sites on the Internet. Our entry into the wonderland of cyberspace is particularly comparable to our broader American experience because it is like the unexpected appearance of a wondrous New World. Space itself has expanded. Umberto Eco called the new electronic cosmos "hyperreality."[3] As for our national geography, Columbus remains a metaphor for all Americans: we still don't comprehend what we have found. Like cyberspace. The challenge has always been not to sleep-walk through our wonderland.

America's First Virtual Reality: *Inventing the New World*

The American novelist F. Scott Fitzgerald recognized the enormous difficulty Europeans faced in identifying the New World: "For the last time man came face to face with something commensurate to his capacity for wonder."[4] At the beginning of the sixteenth century the Atlantic horizon was as blank as the backside of the moon. To fill this void Europeans had long spun fantastic geographies. The mythical Atlantis was a vague and misty land beyond the Western Sea. Antilla was a mysterious place somewhere over the western horizon where Christians had fled from Moslems invading Spain; there they had built seven golden cities (one supposedly located on the site of today's town of Liberal, Kansas).

Europeans did not discover America; they invented it. Strong European prejudices distorted the encounter. Few of their images matched the real thing. Their heads were already filled with dreams of mastering savages and wilderness. Their filters included imperialism, the Christian missionary impulse, a Renaissance compulsion for personal glory, a warlike spirit tending toward butchery, a bundle of old myths about what lay beyond the Western Sea, and greed for gold and spices. Europeans also carried with them their long-standing habit of not investing much value in nature, beginning with the spirit-flesh dualism of Greek philosophy; the Christian belief that man, made in the image of God, was far superior to a soulless nature; and the inflated Humanism of the Renaissance, which sought to appropriate nature through art and science. These filters shut out alternative views and prevented Europeans from seeing the New World on its own terms.

Yet there remained an indisputable reality—a concrete New World geography—beyond the myths, the misunderstandings, and the distorted exploitation. Only reluctantly did Europeans admit to the unexpected existence of a shockingly enormous freestanding continent. The first explorers encountered real rocky shores in the wrong places; a never-ending forest wilderness instead of cities of gold; and strange but tangible plants, animals, and native peoples instead of placid welcoming Chinese in flowing robes. The New World was startling, a physical reality out of nowhere, an astonishing place beyond the wildest fictions. John Smith was not writing fantasy when he described New England streams that really did boil with smelts, salmon, shad, and alewives, while the coast offered lobsters weighing twenty-five pounds.[5] Europeans concluded that the New World was a rich and empty continent, ready to be licked into any shape they decided. They responded sympathetically neither to the fresh treasures of New World ecosystems nor to its successful Indian societies. America never enjoyed its own environmental identity.

After the Explorers, Adventurers, and Settlers
Came the Tourists: *Antiquing through America*

When the English novelist Charles Dickens visited the United States in 1842, he wrote a friend that the reality of America failed to measure up to the mental map he had brought with him: "This is not the Republic I came to see. This is not the Republic of my imagination." On his way across the Alleghenies to Pittsburgh he complained that instead of "scenery" all he saw were "fallen trees mouldering away in stagnant water and decaying vegetable matter, and heaps of timber in every aspect of decay and utter ruin."[6] America was a confusing and disturbing place. When he headed into the Midwest, Dickens added, the flat American prairie left him stripped of words. It was simply wrong. Similarly, the English arch-critic of the 1830s, Frances Trollope, said that beyond the Mississippi River lay nonsense. As late as 1905 the painter Wyrthington Whittredge wrote about the poverty of true landscape in America; we had too much chaos in contrast to Europe: "I was in despair. . . . The forest was a mass of decaying logs and tangled brush wood, no peasants to pick up every vestige of fallen sticks to burn in their miserable huts, no well-ordered forests, nothing but the primitive woods with their solemn silence reigning everywhere."[7]

When Americans say they are "going antiquing," they mean that they are searching out old treasures amid a myriad of junk. Items discarded as worn out or obsolete become cherished for the memories they evoke of an earlier idealized world of handcraft, homespun lifestyle, and rural society. We carry with us a mental picture of what we are looking for, a virtual reality inside our heads. We have in mind a specific oak cabinet or old door lock that has a distinctive shape, texture, color, and dimension. We seek to match these parallel worlds. We search through the vast outside world, having previously invented in our heads what is treasure and what is junk. The actual object turns the mental picture into a physical reality; it corresponds with what we pictured. With luck the newly found treasure of oak or iron also has high intrinsic quality and enduring value. So it is with America's landscape—the nation is one vast antique shop that is scoured for riches by its citizens.

Antiquing in the landscape is not trivial. We have a personal urge to push aside the debris to find the hidden prize. According to the philosopher Mircea Eliade, this includes the urge to be faithfully rooted in the landscape. We want to become safely centered in a physical world and secured to tradition. Such centering especially means awareness of the "other world" that makes the natural or historic site such a prize: it is entry into the "other world" of a superhuman, transcendent plane, the place of absolute realities, giving "birth to the idea that something really exists . . . absolute values capable of guiding man and giving meaning to human existence."[8] Thus, Antique America is not merely nostalgia or romanticism. In it Americans seek, in Aldo Leopold's down-to-earth words, a "base-datum of normality."[9] Space, more than time, roots the American experience; space is the central fact of American history.

The gloomy sights of Dickens, Trollope, and Whittredge are in marked contrast to those perceived by viewers ready to see things that surpassed their dreams. Ralph Waldo Emerson urged his adoring audiences to be like his intrepid "transparent eyeball." What looked like barren ground would be transformed, especially if they would "go out to walk with a painter . . . [to] see for the first time groups, colors, clouds, and keepings."[10] The New York writer Catharine Sedgwick wrote her niece in 1854 to describe how mental pictures fostered real worlds: "I would give a great deal to transfer to you the pictures in my mind of Western life, Western cities, illimitable prairies, and those beautiful, untrodden shores of the Upper Mississippi." Travel made her proud to be an American: "Seeing is believing—the great Valley of the Mississippi, and measuring by that 'the West' beyond." She concluded that she lived in a country superior to the ruined Europe: "I should even venture to put our cheerful dwellings, and fruitful fields, and blooming gardens against the ivy-mantled towers and blasted oaks of older regions, and busy hands and active minds against the 'spectres that sit and sigh' amid their ruins." Earlier, on her way through the Hudson Valley to Niagara Falls, Sedgwick wrote, "This beautiful country stimulates my patriotism."[11] No scenery was more picturesque than the Hudson Valley, no sight more sublime than Niagara.

Antiquing and tourism can also be gullible, shallow, filled with easy stereotypes, and inauthentic—like a visit to the souvenir shops in the Rocky Mountains' Estes Park or to the boutiques at Fisherman's Wharf in San Francisco. The powerful need for authenticity can be misdirected toward fakery in Disney's Matterhorn climb or Mississippi Riverboat trip.[12] By the 1980s the Hershey chocolate show was a fake factory designed for tourists.

Our sleepwalking continues. This led commentator Phil Tippett to conclude that many of America's most-visited places—like Disney enterprises and halls of fame—are ephemera and therefore not trustworthy. Nowhere is the confusion better demonstrated than in the deliberate and convenient re-creations—concrete virtual realities—of Egyptian pyramids and the Italianate Bellagio in the Las Vegas desert.[13] Tippett offers the next step beyond Las Vegas: "The entertainment racket itself is moving towards the idea of virtual reality. What does this mean for Disney and Universal when all you need is a headset and a closet to create an environment, a virtual community? Do they need to keep building these multibillion-dollar theme parks when all of this can be done in a little black box in our heads?"[14] Such a virtual reality swallows up its creators and betrays its confused paying customers.

The History of Mental Mapping in America

A month after the September 11, 2001, attacks on the World Trade Center and the Pentagon the travel section of the *New York Times* invited its readers to reconnect themselves to

America's strengths by visiting its physical icons: California's Golden Gate Bridge; Washington DC's National Mall, stretching from the Lincoln Memorial to the Capitol Building; Chicago's Wrigley Field; the Grand Canyon; Nashville and the Mississippi Delta; Times Square and Central Park in New York City; Iowa's farm towns; Old North Church in Boston; Hollywood Boulevard in California; the Kennedy Space Center at Florida's Cape Canaveral; and Franklin and Eleanor Roosevelt's Hyde Park along the Hudson River. The *Times* described these "American Treasures" as "grand and simple symbols of the country [that] stand ready to revive spirits."[15]

American travelers have for decades crisscrossed the land-scape in search of a national identity. They have found it not only in Niagara Falls and the White Mountains but also at historic sites—military installations, battlefields, cemeteries, hallowed buildings, and historic markers.[16] Yosemite and Yellowstone provide splendid scenery sublimely set aside in jewellike national parks. Ghost forests such as those holding the remaining California coastal redwoods are still haunted by the specters of great, towering trees sacrificed for profit and lost forever. At man-made historic sites Americans see themselves as heirs to Mount Vernon, Lewis and Clark's Fort Clatsop, and the Gettysburg battlefield. Some Civil War battlefields for years exposed "skulls, arms, legs, etc., all bleaching in the sun." Contrived presidential faces were carved on Mount Rushmore, in South Dakota's Black Hills. In the twentieth century such shrines came to include presidential museums; Elvis Presley's Graceland in Tennessee (reputedly the most visited site in America); and halls of fame celebrating football, baseball, rock music, and NASCAR. Western ghost towns are haunted, like California's forests, by the specters of "genuine Americans"—the miners, cowboys, farmers, outlaws, and federal marshals of an earlier, simpler day. Thus, some American treasures are natural, like the Smoky Mountains and the Grand

Canyon; others are man-made, like the Brooklyn Bridge, the Saint Louis Arch, and—dare we say?—Las Vegas. Such sites have long enabled tourists to locate themselves not only in America's perfect space but in America's perfect time as well.[17]

These searchers also found "Authentic America" in its emerging industrial might. Despite their grit and grime, factory operations glorified American achievement and industriousness. By the 1820s tourists strolled through Mauch Chunk, on the edge of Pennsylvania's anthracite coalfields, as well as Lowell, Massachusetts, home of the revolutionary textile mills of New England. By the 1830s they peered into Pittsburgh's ironworks, its glassworks, and the digs in its Coal Mountain. After the Civil War Chicago's Union Stock Yards were a major tourist destination, as Ford production was in Detroit in the twentieth century.[18] One could see corn flakes poured into Kellogg's cereal boxes in Battle Creek, or pickles bottled into Heinz glass jars in Pittsburgh. Braver folk perched themselves above roaring steel mills in Gary, Indiana.

World's fairs clustered all the inventive wonders together. Historian Cindy S. Aron describes visits to the Philadelphia Centennial Exposition: "Those who stopped to contemplate the gigantic Corliss engine at the 1876 Philadelphia fair might well have felt a kind of awe similar to that experienced as Niagara, the former offering evidence of the power of man while the latter suggested the power of the divine." She adds: "What a visitor to a world's fair could not encounter were the scenic vistas and natural wonders of the American landscape—highpoints on most tourists' journeys. But for many . . . the fair became a sight in itself, making its own landscape and producing its own wonders."[19] World's fairs not only celebrated the nation's future but concentrated the entire world in one place at one time for public inspection.

However, growing disillusionment with American industrial society eventually increased interest in nature's wonders. Such disillusionment goes back to Jefferson and Thoreau, Olmstead and Muir, and includes the "back-to-the-country" Garden City movements and the general exodus to suburbia. One peak time for antiquing outside factory cities came during the Great Depression of the 1930s, when many Americans sought reassurance in the Great American Outdoors as they found flaws in the industrial system. Despite the poor economy, tourism boomed in the 1930s.[20] As a 1935 editorial in the *New York Times* put it, travel was a form of pilgrimage that allowed one "to look into things but never merely to look at them."[21] Ritual visitations to the Grand Canyon, for example, increased by one-third in 1934. One suburban New York woman deliberately sought simplicity in a cottage on the Carolina coast, citing "blue sky, yellow sand, the music of the surf . . . good food, Southern talk," although she also mentioned "all modern conveniences."[22] Camping and hiking took off in the Adirondacks. Much of the make-work that employed the young men of the Civilian Conservation Corps (CCC) focused on bridges, roads, and buildings in the national parks and other public lands.

Arcadia: *The Gift of Rural Antiquing*

In the early nineteenth century, when American travelers began to tour their own land, they did not seek out rugged and uncleared places. Instead the early tourists delighted in cultivated land, where farmers could be seen as coworkers equal to God. Americans located Arcadia, the classic middle landscape of Ancient Greece, in their own landscape. The Arcadians, relatively isolated like Appalachian farmers, had lived a simple and natural life. As early as the 1780s Crèvecoeur and Jefferson heaped praises upon the yeoman farmer, identified as the truest American. Such icons cleared out the wilderness, established the rural landscape, and ordered the land. A century later, in the industrialized 1880s, the influential essayist Horace

Bushnell could still praise rural America as "a second creation at the hand of man . . . added to the rudimental beauty of the world."[23] In the 1920s the European visitor Count Keyserling described a Kansas landscape as nearly perfect because it was the most fully domesticated place he had ever seen.[24]

Arcadia's dreamworld-come-true offered personal renewal. The search for "just the right experience" induced adults to spend their vacation time briefly boarding and working on a farm—the "hills of home" in Vermont, New Hampshire, Maine, or western Massachusetts. One city businessman told of rediscovering "boyish sleep, appetite, and afternoon staying power." Nor did he mind putting unused muscles and unstretched tendons to use: "Blisters—blisters?—a spot without them became notable." He also knew he would walk away from the farm and return to his comfortable office. In the meantime he could feast on "the sweetest butter, the freshest milk, the purest maple syrup." He could visit the little red schoolhouse, go to Sunday church services, hear the band play in the town square, and wander dreamily along the local stream. Governor Frank Rollins of New Hampshire effused, "When you think of the old home, you bring back the tenderest memories possessed by man,—true love, perfect faith, holy reverence, high ambitions—the 'long, long thoughts of youth.'"[25] The summer visit memorialized "Authentic America." It harbored nicely simplified values of the "home place"—peace and contentment, community cohesiveness, and republican simplicity restored against modern greed and crude competition.

Entire regions, like New England, spotted with white-painted, church-steepled towns surrounded by picturesque small farms, became iconic entries into a historic dreamworld. Wily local promoters turned dilapidated town centers into "old Northhampton" or "mysterious Salem." Martha's Vineyard became a rustic throwback to an earlier era instead of a dying outpost, while Nantucket Island's struggling whaling industry appealed to the sentimental imagination.[26] Later both became escapist havens for prosperous retirees. In the words of historian Dona Brown, "New England's countryside was imagined as a kind of underground cultural aquifer that fed the nation's springs of political courage, personal independence, and old-fashioned virtue." The outcome, says Brown, offered Americans what they needed, "a mythic region called Old New England . . . a reverse image of all that was most unsettling in late nineteenth-century urban life."[27] This quest was backward-looking, from modernity toward an American never-never land that in the mind's eye blended beauty, harmony, order, and peace.

Antiquers rejoiced over a trove of regional treasures. The Adirondacks, the nation's largest protected park, held a maze of lakes allowing the New Yorker to return to a primeval privacy. Americans sought secrets to authentic America in Appalachia because the region seemed frozen in colonial time, with writing and music that preserved Elizabethan and Scots-Irish settlement. The South was imaginatively transformed into the Old South before the Civil War, a plantation world of genteel women, dashing men, childlike "darkies," and grand moss-covered trees. There were also literary sites such as Mark Twain's Mississippi River, where the archetypal Huck Finn discovered himself, and Hemingway's northern Michigan, where Nick Adams evoked a classic American coming-of-age. Willa Cather, in her description of Nebraska frontier grassland, bonded immigrant dreams with purification earned out of the difficult Plains. Regional artists, such as Thomas Hart Benton, John Stuart Curry, and Grant Wood, stylized a perpetual American heartland that would have pleased Crèvecoeur and Jefferson. One of the most intriguing restatements, entirely depending upon regional images, came in journalist Joel Garreau's 1981 *Nine Nations of North America*, in which he ignored state lines and national boundaries to divide the continent into New England, the Foundry (the mid-Atlantic and the Midwest), Dixie, the

Breadbasket (the Midwestern corn belt and the Plains wheat belt), Mexamerica, Ecotopia (northern California and the Pacific Northwest), the Empty Quarter (from the Great Basin north through the Canadian Rockies), a free-standing Quebec, and the Islands (the Caribbean).[28] Garreau urged Americans to acknowledge regional myths as geographic realities.

Viewing Nature's Nation

In 1837 over three thousand people visited the wonder of the Connecticut Oxbow, connecting the view to what they had already seen in a large landscape painting by the English expatriate Thomas Cole. In 1836 Cole had exhibited *The Oxbow*, which depicted on the left side a rugged and tangled wilderness and on the right side a calm agricultural landscape. This was a window through which one could glimpse the meaning of America: we peer from a civilized plain into the heart of nature. One contemporary observer, Rufus Choate, wrote that landscape painting combined "telescope, microscope, and kaleidoscope all in one."[29] With a telescope one scanned the distant landscape to find a national identity; the microscope zeroed in on the bond between self and wilderness; turning the kaleidoscope gave the viewer the dazzling array of the natural environment. Scientific-exploration and railroad-survey teams took along renowned artists. When Thomas Cole, Albert Bierstadt, Frederick Church, or Thomas Moran exhibited a new canvas in an Eastern gallery, it was a major public event that drew large crowds and newspaper headlines.

These landscape artists promoted American scenery as the transcript of national character.[30] They gave Americans a visual language—a set of reference points—about the nation's geography. Long before the English religious poet John Donne had described such images as "Nature's Secretary," since nature itself is mute. Cole set a standard when he offered the opinion

7. Thomas Cole, like Albert Bierstadt, became famous for his mythical or invented scenes. The Oxbow of the Connecticut River, however, was a real place that we can still find today, just as we can still locate the sites depicted in Bierstadt's images of Yosemite. Cole's painting informs Americans of the contrast between rural and pastoral Arcadia on the right side and a rough, threatening, but not entirely dangerous wilderness on the left side. Cole suggests that both are valid depictions of American grandeur; the two American realities should be seen as both complementary and competitive. Cole's painting of the Oxbow immediately drew tourists to compare the real geography with the painted interpretation. Some might have concluded that the painting was more truthful than the real thing.

that landscape painting contained a patriotic message: "Nature is the national past, the basis of the national identity, an infinite source of moral regeneration, and guarantee of the democratic constitution." The artist sought to make the viewer proud of an imagined American wilderness—virtual America became authentic America. The natural landscape provided the history and culture that Americans felt their nation lacked when they compared it to the Old World with its rich past. Americans looked at their landscapes (through artists' eyes) and concluded that they were truly Nature's Nation. Landscape painting was not simply artistic: it had a public function; it created "the iconography of nationalism." Art historian Barbara Novak writes that landscape artists "were among the spiritual leaders of America's flock."[31]

Wilderness antiquing became a religious pilgrimage. The public believed that landscape artists gave them truthful details of a vast American panorama, felicitously combined with "beautiful sentiment." Church, Moran, Cole, Bierstadt—all created visual scripture that told of sacred places that actually existed in the public domain: Niagara Falls, Yellowstone, Yosemite Valley, and the Grand Canyon. There were even groves of sacred trees like those of the ancient world; in America's case they were enormous redwoods or sequoias. These were scenes of holy power, displays of nature's force. This wilderness was primeval, archaic, and even rawly primitive—the way things were before being spoiled by human greed and indifference. Original America promised an endless landscape of nature's plenitude, abundance and resilience. (Native America was deliberately ignored.)

The landscape artists were one with the romantic, nature-loving Transcendentalists. Emerson wanted nature and nature's God to enfold him in a mystical union. Thoreau's message from Walden Pond is still attractive in its rebellious invitation to escape mechanical and commercial civilization in order to be immersed in nature, if only for a few days or weeks. This was life stripped down to its essentials. In art historian Angela Miller's words, "Thoreau reminds us that only an identity rooted in the soil of place, particularized memories, and personal associations can keep us true to ourselves and our communities, allowing us to judge national objectives with greater clarity and skepticism."[32] Thoreau's 1846 travels into the wild heart of Maine pushed him to scribble a note: "Think of our life in nature,—daily to be shown matter, to come in contact with it,—rocks, trees, wind on our cheeks! The *solid* earth! The *actual* world! The *common sense*! *Contact*! *Contact*!" Literary critic Wright

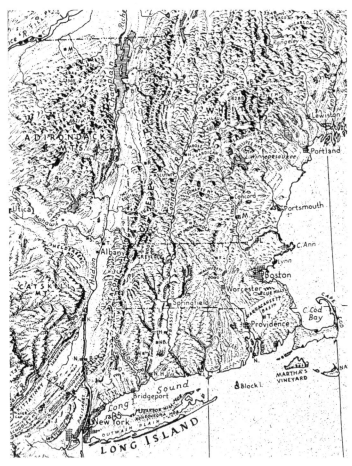

8. Erwin Raisz's hand-drawn physiographic maps from the 1940s and 1950s are still exemplars of their craft. In this detail from a large U.S. map we can follow the route of the first major tourist trek, in the 1830s, from New York City: along the Hudson River following the Palisades to the Catskills and Adirondacks (Mount Marcy) along the left side, then into the Green and White mountains (Mount Washington) on the right side. The Connecticut River of Cole's Ox-bow marks the division between the Green and the White mountains. Tourists looked upon this Grand Tour as equivalent to the Englishman's Grand Tour of the antiquities of Europe. Later, after the Civil War and with more ambitious tours to Niagara Falls, Virginia's Natural Bridge, and the West's Yosemite and Yellowstone, Americans concluded that their landscape contained sights superior to anything Europe offered.

Morris concluded that the Transcendentalists provided "the first contour map of what we might call our *natural* state of mind." Later, in 1903, John Muir, as heir to Emerson and Thoreau, stated an American affirmation of wilderness: "Thousands of tired, nerve-shaken, overcivilized people are beginning to find out that going to the mountains is going home; that wilderness is a necessity; and that mountain parks and reservations are useful not only as fountains of timber and irrigating rivers, but as fountains of life."[33]

How did Americans do their antiquing? How did they travel, what technologies did they acquire to "see," and what did they find?

Virtual Reality Takes Hold: *The Grand Tour*

By the 1830s middle-and upper-class Americans, well informed by artists and guidebooks, took their own version of the English gentleman's European Grand Tour. By boat and stagecoach (and soon railroad) travelers worked their way up the Hudson to Albany, then to Lake George and the Berkshires, Mount Holyoke and the Connecticut Oxbow, and finally New Hampshire and the White Mountains. In the process they experienced all the requisite iconic scenes—rivers, waterfalls, lakes, mountain vistas, dark forests, and pastoral landscapes. The Hudson was America's Rhine. Lake George served as Lake Como. New Hampshire was America's Switzerland, and the White Mountains did for the Alps. Imitation was deliberate: Americans felt culturally barren until they could match European vistas. The tour also included European-style health spas such as Saratoga Springs and White Sulphur Springs, as well as grand resorts at Lake George.

Guidebooks directed the traveler to the Grand Tour's high points. Americans learned to read landscapes as they might read a book. Indeed, wilderness was America's scripture. Guidebooks

9. The emerging network of railroads offered the first wide public access to America's great natural places. This photo of the Haines Corners station seems to have been taken about 1900, judging from the women's clothing and the children's pose. The railroad prospered by guaranteeing "prompt connections by boat or rail" deep into the Catskills. The summer-only Kaaterskill Railway, an underpowered, narrow-gauge line, anticipated the classic railroad-resort connection, in this case to the popular Catskill Mountain House (built in 1823, open until 1941, and burned down in 1963). The mountainous terrain kept train speeds below what automobiles would soon accomplish.

emptied information and images into tourists' heads. Tourists expected to see a panorama of European-like wonders and American curiosities. *Putnam's Monthly Magazine* told its readers in 1854 that the wilds of northern New York offered not only pure air and fresh breezes but also sublime scenery that would restore the intellect and imagination. The popular *Appleton's Illustrated Handbook of American Travel* guided 1857 travelers to cities, towns, waterfalls, battlefields, mountains, rivers, lakes, hunting and fishing grounds, watering places, and summer resorts.[34] People went to see what they already knew was there; they verified what the guidebook recommended. Guidebooks told visitors to Colorado's Garden of the Gods which rock formations represented castles and which represented humans.

10. The four main transcontinental railroad lines that crossed the Rocky Mountains led their magnates to discover another source of wealth in vacation resorts. As this photo demonstrates, American tourists were not interested in roughing it, but insisted on comfort, even luxuries, and an easy walk or ride to an "inspiration point." Manitou Springs Onaledge, continuously in operation over the last century, provided access to one of the most challenging ventures: the top of Colorado's 14,110-foot Pike's Peak—first by an 1891 cog railway, then by a 1916 auto roadway. Great lodges opened the doors into Glacier, Yellowstone, Grand Canyon, and Yosemite national parks. In some cases the railroads, by creating access to these natural theme parks, can be said to have invented them. Railroad tours to the national parks lasted into the 1950s.

The result was a mental list of icons that shaped tourists' experience of the real places. Indeed, guidebooks claimed to be so accurate that readers didn't have to actually observe every site for themselves.

The tour, however, was not taken out of a general liking for the outdoors. Rather than venturing into the actual messy nature of real trees, rocks, mud, rain, and steep paths, many Americans decided that it was better to do their aesthetic and patriotic fieldwork in the local museum or gallery. Art historian Angela Miller writes, "Americans' vaunted love of nature proved to be . . . better served by images than by the thing itself."[35] Americans did not want nature on its own terms. Even Thoreau's Walden Pond was free of dangerous animals and Indians, and town was but a short walk away. The passage westward by train across the oceanlike Great Plains was to be rapid, comfortable, and convenient. The tourist's vacation was intended to produce a change of pace and a change of scene, but not hardship or privation.

Today's antiquers look not only for scenes described in guidebooks but also for those selected by motion pictures, seen on television, searched for on the Internet, depicted in advertisements and articles in newspapers and magazines, reintroduced by visits to galleries and museums, learned of from the advice of friends, and remembered from school geography lessons. At worst the antiquer uncovers only "pseudo-events," in historian Daniel J. Boorstin's words. The modern American tourist, Boorstin says, "has come to believe that he can have a lifetime of adventure in two weeks. . . . He expects that the exotic and the familiar can be made to order. . . . Having paid for it, he likes to think he has got his money's worth."[36]

Master Scenes and Inspiration Points

A layering of filters on wilderness created for Americans a comprehensive mental picture of ideal landscapes. As early as 1790 the Englishman Archibald Alison told Americans that a scene had no intrinsic attraction but pleased the viewer because of the trains of thought or "associations" that it set off in the viewer's mind.[37] Landscape artists and writers, and after 1850 photographers, selected details of the environment and out of them synthesized a highly interpreted panorama. Looking at Grand Teton, and looking at a painting of Grand Teton, involved a feedback system, a circle of constant comparison, discovery, and modification. The tourist looked for a series of

11. This large landscape painting serves mental expectations (and creates them) on several levels. In the background is California's iconic Yosemite Valley, emerging in the 1870s as America's representative natural place, superseding the Catskills, Adirondacks, and Niagara Falls. Yosemite, made known by Frederick Law Olmstead and soon home to John Muir, would not become a national park until 1890. In the foreground are overdressed Victorian tourists, men and women, who reached the vista via the relative comfort of horseback. The site itself was perhaps America's best-known wilderness "inspiration point": Glacier Point. Artist William Hahn put the viewer above the tourists, the point, and the valley, granting a stupendous sight. Not least, this painting may be the only large Yosemite canvas that includes human figures; the traditional Yosemite landscape by Bierstadt, or photograph by Eadweard Muybridge, offered "pure" wilderness, consciously or unconsciously filtered by the painter or photographer. Hahn, German-born and German-trained (like Bierstadt), was a successful landscapist before emigrating and opening a studio in San Francisco in the early 1870s. He carried European landscape conventions with him into his American art.

"master scenes," indicated by the requisite stops at "inspiration points" that matched the Currier & Ives print, the guidebook, or the photograph. The master scenes at Yellowstone included the geyser basin, Mammoth Hot Springs, and the Canyon of the Yellowstone. At Yosemite a single vista encompassed Half Dome and Bridal Veil Falls. But in both parks the rest of the wilderness held for tourists only dreariness and disappointment. (The Greater Yellowstone Ecosystem concept would have to wait almost another century.) Both parks had originally been established to preserve "natural curiosities," not representative wilderness. The modern parallel is, of course, the two-week automobile vacation, covering six thousand miles and twelve national parks and monuments and documented by ten rolls of color film or five hundred digital images. Today's traveler comes upon the scenic turnout or viewpoint, already chosen for his or her benefit. The window to the world, once in the parlor, is now the auto windshield.

Envisioning Technologies

The Claude Glass

French landscape artist George Claude urged tourists to view their scenes through a sepia-toned looking glass, which came to be known as the "Claude glass." Such viewing would enhance the natural scene and give it a painterly quality. The Claude glass brought on a plethora of pink, red, and rust wilderness-sunset canvases. The viewer literally looked at a pleasant scene "through a rose-colored glass." The popular impact of this technology is suggested by the ubiquity of the sepia-toned prints that dominated the early era of commercial photography.

The Mechanical Panorama Canvas

Nineteenth-century Americans rushed in droves to see the moving panorama, a long, elaborately painted, bannerlike reel of canvas, about five feet tall, mounted on rollers that moved the scenes across a stage in front of the audience. Historian Anne Farrar Hyde concludes that the panorama "created the closest illusion of reality that audiences had yet seen." The most famous canvas was John Banvard's *Panorama of the Mississippi River*, in which the stage became a riverboat deck, while the painted scenery scrolled along the back wall, to create a vivid journey aboard ship. Banvard's advertising claimed that his stupendous canvas was painted on "three miles of Canvas; Exhibiting a View of Country 1200 Miles in Length, extending from the Mouth of the Missouri River to the City of New Orleans."[38] Banvard boasted that four hundred thousand Americans had taken this journey in the first two years it was offered. He deliberately emphasized sublime mountain torrents, European castle–like bluffs, and picturesque details of everyday life along the river.

Other panoramas informed audiences about Western adventures such as the California Gold Rush. James F. Wilkins offered audiences the vicarious experience of Westering in his *Moving Mirror of the Overland Trail*. One enthusiastic reviewer described the iconic panorama: "Mountains with their snowy tops and ragged sides, seen in the distance, gilded by the last rays of the setting sun, remind us of all that Byron or Coleridge have written or sung of the far-famed Alpine scenery."[39] A combination of daguerreotype and painting showed up in the popular *Jones' Pantoscope of California*, which was somewhat more realistic about plains monotony, mountain travail, and dangerous desert crossings. The influential clergyman Henry Ward Beecher wrote of the *Pantoscope*, which was playing to overflowing houses in Boston: "It communicates important knowledge about a large tract of our own territory, the like of which for its peculiar, wild, and original features, is nowhere else to be seen on earth."[40]

The Magic Lantern

Americans soon thrilled over a new technological wonder, magic-lantern shows that used a primitive slide or filmstrip projector

12. Photography, still a young medium, took a very large step by creating the illusion of a three-dimensional image. Families collected stereographic cards, where two identical but offset images were printed side-by-side. The three-dimensional image convinced nineteenth-century Americans that they were breaking the barrier between the reality of the photograph and the reality of the actual scene. A hand-held viewer became a popular household item. Major photographers of the American landscape, such as G. E. Curtis, offered these cards as virtual realities of great American landscapes. Kodachrome color three-dimensional images, appropriately enough, first appeared at the 1939 New York world's fair, in a seven-image View-Master reel. Such images remained popular approaches to American national parks and other sites through the 1950s. And even though three-dimensional images were an early photographic phenomenon, the success of their realistic depiction remains unmatched even today, whether in the movies, on television, or on the computer screen.

to display still pictures upon theater screens. Hyde notes the importance of photography as a new medium but maintains that "despite [photographers'] claims about the truthfulness of photography, their images echoed the tradition developed by explorers, writers, and artists in earlier decades."[41] The illustrated travel talk—the "vicarious tour"—that had begun with the panorama would continue into the era of movies and television. Until the 1960s dominance of television Americans crowded into local auditoriums to see the latest "travel and adventure" movie lectures about Yellowstone or the Grand Canyon. The pioneering magic-lantern lecturer Stephen James Sedgwick showed Americans what they were already prepared to see, and newspapers wrote major reviews of his presentations: "His audiences were transported to distant wilds, to terrific heights, translucent lakes, and natural scenes of . . . peerless subliminity." Another review added: "We were made to feel eager to journey over the route to witness the scenes so charmingly presented."[42] Sedgwick in fact was a lobbyist for the Union Pacific Railroad. Hyde notes that "the images from the magic lantern suggested to Sedgwick's audiences that the Garden of the World did truly exist and that the desert had been conquered."[43] And the Union Pacific promised to carry visitors safely to these exotic locations.

The Stereoscope

Stereographic pictures enhanced Virtual America through their vivid three-dimensional effects, a revolutionary step in making the reality of any scene perceptible. The stereographic photo was constructed of two adjacent paper or glass images of the same scene, slightly shifted, that could be viewed through an inexpensive and convenient handheld device. Hyde emphasizes the fact that "the stereoscope allowed a photograph to fill the viewer's entire field of vision [which] created a startling illusion of reality."[44] The stereoscope brought its three-dimensional version of virtual reality directly into the home parlor by the millions, becoming "the poor man's picture gallery," according to *Frank Leslie's Illustrated Weekly*. The photographer Oliver Wendell Holmes explained this new virtual reality: "The first effect of looking at a good photograph through the stereoscope is a surprise such as no painting every produced." He admitted the deception, the creation of a mental picture: "The mind feels its way into the very depth of the picture. The scraggy branches of a tree in the foreground run out at us as if they would scratch our very eyes out." Holmes even predicted that stereoscopes would make travel unnecessary: "These sights are offered to

13. Americans considered Albert Bierstadt's large landscapes of their scenery to be representative of their natural sublime. This 1868 canvas, painted shortly after the divisive Civil War, suggests a common ground through which all Americans might rediscover their national identity. The site, an imaginary place that includes the different dramatic aspects of the Sierra Nevada, embodies Bierstadt's (and America's) image of the mountains of the West, thrilling viewers and enticing tourists to visit the sacred ground. The most powerful features of the sublime are included, notably a mountain range made mysterious by clouds and storm. The painting includes the requisite dramatic water site, wild greenery, and even undisturbed wild animals. Viewers of the canvas believed they could be personally enriched by the image itself, its own virtual reality.

you for a trifle, to carry home with you, that you may look at them at your leisure, by your fireside."[45] Particularly popular in the 1860s was the series *Yosemite Valley, California,* by photographers E. Anthony and H. T. Anthony. Most of the scenes in this series were dramatic landscapes, with foreground rocks, trees, or standing figures to guarantee the three-dimensional effect. They deliberately reproduced what viewers had already been trained to see by earlier landscape painters and guidebooks, still in black-and-white but now with more dramatic realism. Alternatively, various series of stereo photographs of the transcontinental railroad, gold-mining flumes, disappearing Indians, and burgeoning cities reinforced American bedrock belief in the nation's Manifest Destiny. The stereoscope's popularity revived in the 1950s, featuring color transparencies of America's special places seen through the View-Master reel and viewer, which remained popular for thirty years.

Nineteenth-Century Mental Filters

If nature was a parade of wonders presented for brief inspection, how were these wonders chosen? Landscape artists depended upon three mental templates that they layered onto wild nature to turn it into representative national scenery.

The Sublime

In the middle of the eighteenth century, when America was still British, Edmund Burke, the statesman and philosopher, took a look at nature, centered his attention on wilderness, and called it "sublime" according to a set of particular features: obscurity (darkness or mist), power (storm or waterfall), privation (emptiness and silence), vastness (remoteness), sense of infinity, eternity, difficulty of access or traversal, and magnificence (a dramatic sunset).[46] These elements aroused overpowering human emotions of wonder, respect, and reverence. Burke focused on the impact of the sublime upon the observer's state of mind—for example, when encountering an impassible mountain range. The modern French essayist Alain de Botton identifies the sublime with the discovery that "the universe is mightier than we are, that we are frail and temporary and have no alternative but to accept limitations on our own will; that we must bow to necessities greater than ourselves."[47] The observer did not find the landscape paintings of Bierstadt or Moran restful; they were designed to be exhilarating, cleansing, and redeeming, sending thrilling sensations of awe and dread, and sometimes joy and gladness. Water never simply flowed; it crashed and rushed and roared on the canvas. Trees and plants were

14. English-born, Ohio-bred, and New York–established Thomas Cole combined landscapes and allegories in his large paintings, which drew crowds wherever they were exhibited. Considered one of the founders of the influential Hudson River School of landscape painting, Cole also wrote widely about the connection between America's primeval landscapes and a "true" national identity. While *Falls of the Kaaterskill* has features of the sublime, notably the stormy and dramatic setting, this is overwhelmed by its picturesque features, characterized by broken trees ("blasted trees"), windblown branches, clutter and rubble, both subdued and overwrought colors, and the posed isolated Indian. The artist's vantage point is a puzzle, since it appears to be some dozens of feet into thin air, facing the falls, based on the author's visit to the site.

15. Yosemite Valley became the West's iconic symbol, created by excited written reports and even more by paintings, drawings, engravings, lithographs, and photographs. This trite monographic image is typical of what Americans might have seen in newspapers, illustrated magazines, gushy patriotic tabletop books, or in framed pictures on parlor walls. In this case the influential garden-park style dominates even Yosemite's sheer walls: a flat grassy meadow, an ambiguous waterway, and clumps of trees and shrubbery, all dominated by the requisite S-curve pathway inviting nothing more strenuous than a stroll. There are small touches of the sublime in the distance and the picturesque in the foreground, but this landscape is neither dramatic nor threatening. Indeed, the pathway curves into the valley, inviting the visitor to wander further instead of standing struck by wonder. Americans were particularly pleased that the garden park, so assiduously contrived in England, seemed to appear naturally in their wild places. The Old World's simulation was the New World's reality.

always primal, "animated by the breath of life," and heavily fecund. The colors were always blacks, browns, dark greens, and hazy yellows—not bright pretty reds, whites, and blues. Nature's sublime induced awe and a sense of the holy. Americans quickly attached the sublime to the American landscape, particularly its qualities of vastness, infinity, and magnificence. A visit to a wilderness park like the Adirondacks or Yosemite left "marks on the mind."

Ideally, the experience of the sublime was literal, direct, and immediate. Jefferson in 1786 described the dramatic view from the heights of Monticello to his bosom friend the landscape artist Maria Cosway: "Where has nature spread so rich a mantle under the eye? Mountains, forests, rocks, rivers. With what majesty do we there ride above the storms! How sublime to look down into the workhouse of nature, to see her clouds, hail, snow, rain, thunder, all fabricated at our feet! And the glorious sun, when rising as if out of a distant water, just gilding the tops of the mountains, and giving life to all nature!"[48] The government surveyor Clarence King, in 1871, climbed California's Mount Tyndall, there to encounter, he said, terrible desolation and intense self-awareness: "Nature impresses me as the ruins of some bygone geological period and no part of the present order—like a specimen of chaos which has defied the finishing hand of time." The physical confrontation morphed into a spiritual encounter. On the mountain, which was a sublime expression of nature, exclaimed King, he was able to define his personal place in the vast scheme of creation: "Rising on the other side, cliff above cliff, precipice piled upon precipice, rock over rock, up against the sky towered the most gigantic mountain wall in America. I looked at it as one contemplating the purpose of his life."[49] Setting aside the Christian self, and the rational self of the Enlightenment, King believed he had merged himself with a national spirit that sprang forth from America's geography.

The Picturesque

Like most good ideas, Burke's sublime was corrupted—by William Gilpin's "picturesque." Gilpin delighted the public by adding bric-a-brac that soon became requisite landscape furniture: rocky chasms, blasted trees, rustic bridges, even herdsmen and lowing kine. The end product was a softened and accessible wilderness or a gone-to-seed rural countryside. The packaging of the visual experience may have reached self-caricature with the picturesque. The famous Catskills resort Mohonk Mountain House, still in active use, is a combination of fortress and rustic mansion, its gentle paths around the lake studded with rough-hewn benches and gazebos placed at "inspiration points." For Americans the classic picturesque image was likely to include a solitary Indian pondering a waterfall in the Catskills. The picturesque would soon collapse into the trivial prettiness of Victorian sentimentalism.

The Garden Park

To these touristic images we must add a third mental template: the English landscape planner Capability Brown's "garden park" scenery. Its markers were a grassy open field with a me-andering stream or path in the requisite S shape, dotted with separate clumps of trees or bushes. The garden park, or bucol-ic rural setting, was likely to be presented as a vista framed by dark masses of green foliage, the entire setting centered upon a prominent body of water, the atmosphere fuzzy with mois-ture. The scene would emphasize verdant growth, even rotting fecundity, implying layers of squandered abundance far beyond ordinary human needs.

The image became the model for American city parks de-signed by Frederick Law Olmstead and Calvert Vaux, as well as the lawns of middle-class homes. Its familiarity made high meadows in the Rockies attractive to tourists in places with names like Estes Park and Winter Park. We return to Arcadia.

Antiquing the West: *"The Best Is in the West!"*

The American West stood out as the premier region for an-tiquing. Tourists looked to the West for an "Original America" that had sprung from the untamed land inhabited by untamed people. This was no ordinary wilderness, but a geography over-

16. Yellowstone Inn, completed in 1904 as a tourist centerpiece for Jay Cook's Northern Pacific Railroad's development of Yellowstone National Park, cost the railroad two hundred thousand dollars. Historian Alfred Runte claims that "the railroad's dependence on unspoiled scenery to attract tourists tempered its purely extractive aims, such as logging, mining, and land development." By the 1920s the combination of the automobile and the West's natural wonders was bringing families by car in record numbers. After World War II Yellowstone and other parks were invaded by millions of families in their cars, changing the parks forever, but not the image first promoted by the railroads.

whelmingly superior to Europe's mere sublime. As early as the 1830s antiquers were heading west, not far behind government expeditions and immigrant wagon trains. Railroads and steamboats soon opened the West, first for the well-heeled few and soon for the middle-class "vacationer" who lingered at the vast wooden hotels and lodges of Wyoming's Yellowstone Inn, Colorado's Manitou Springs and Glenwood Springs, and El Tovar at the South Rim of the Grand Canyon—lazy and relaxed, loafing on porches punctuated by leisurely walks to nearby vistas. Pilgrims admired the peaks of the Rocky Mountains and wandered through the bare pinnacles of Colorado's Garden of the Gods. This style continued into California resorts at Monterey's Hotel Del Monte, the Hotel Del Coronado at San Diego, and Santa Catalina Island.[50]

Antiquers "collected" experiences in their diaries recounting famous sites like shopping lists. Pilgrims were delighted to track down physical places that "proved" the reality of a mythic West that otherwise existed only on the great landscape canvases, on the pages of guidebooks, or in American brainpans. There were superpowerful places like Mount Rushmore or Wounded Knee; extraordinary heroes like Jedediah Smith or the outlaw James Brothers; crucial sequences of events like the Indian Wars, the capture of territories from the Spanish and Mexicans, and the building of the transcontinental railroad. Of Yellowstone travel writer Frederic Van de Water observed that "parties would come in the East Gate, drive directly to Old Faithful; see it spout and then go out the West [Gate]. That is the sole thing at which many want to look. There were pictures of it in their geography books."[51] Charles Finger wrote in 1932: "We boiled an egg in the Frying Pan Hot Spring, watched artists painting the Yellowstone Canyon . . . rowed on Yellowstone Lake, saw Old Faithful erupt three times, admired Morning Glory Pool . . . [and] witnessed an eruption of the Lioness Geyser, which is a rare sight."[52] The essayist Melville Ferguson described a landscape constructed for the tourist—walkways amid the geyser basin, grand hotels with viewing balconies, all like a city playground.[53]

The West's extraordinary settings of mountains and deserts, wagon trails and battlefields, became sacred, a space where Americans could find validation of their individual beings and their national destiny. Van de Water wrote in 1927, "Here [in Yosemite], something whispers, is one small rectangle [of] the world where life goes on serenely, tolerantly as the Architect planned it before Eden."[54] Long ago" was shrouded in mystery; in geographer Yi-Fu Tuan's words, antiquity was "the time when the gods still walked the earth, when men were heroes and bearers of culture, and when sickness and old age were unknown."[55] The West stood out as a sacred terrain that opened doors leading toward the Holy, the Transcendent, and the Triumphant. The West unveiled the mystery of what it means to be an American, its effect similar to that of the pilgrimages undertaken by religious devotees worldwide to Lourdes, Rome, Mecca, and Mount Fuji. Timeless Edens and utopias lay in distant Western places.

The young Wisconsin historian Frederick Jackson Turner tapped into this mythic West: his fabled "frontier thesis" gained its own life as a virtual reality. First in 1893, Turner described the ever-receding West as the source of self-reliant individualism, democratic institutions, family-centered communities, and optimism about the future. Turner and his heirs told Americans that this West defined how a people had come into existence, what shaped their primary characteristics, and what foretold their significance in the course of human events. Twentieth-century tourists also sought out a "John Wayne" mentality in the West: a revolt against urban modernity and a hankering after a history filled with simple values, well-defined heroes, and recognizable guideposts.[56] This was a traditionalism that proposed that the nation could not let loose its frontier heritage except at its peril. Today the West remains deeply entombed in Turner's words and imagery.[57]

Virtual Places: *America Deserta Shifts from Empty Space to Sacred Place*

No place was more off-limits than the Southwestern deserts. Biblical tradition had identified desert regions as places of evil and primeval chaos. The cursed Sinai had to be painfully crossed before the Promised Land, Israel, could be entered. A desert may in truth be ecologically rich and varied, but the leg-

17A and 17B. Shiprock, a volcanic plug near the Four Corners region (a), and a desert scene in southeastern Utah (b).

endary image of the desert presented an environment that is utterly hostile, empty of shelter and life support. Psychologically, entering the desert courted madness and the dissolution of one's being in a primeval chaos. Explorers like Zebulon Pike and John C. Frémont reinforced this image of wasteland, reporting the persistent desolation of the Great American Desert. Desert travel involved extreme physical discomfort and likely death forced by thirst and heat. Immigrants to California dreaded the passage through Utah and Nevada. A local horseman accompanying a geologist into Utah's Canyonlands in 1936 simply said, "There's just gotta be mineral in those rocks; no piece of country could be so gosh-darned [sic] worthless."[58]

The landscape painters who had so dramatically shaped virtual realities in American heads openly repudiated desert regions. In 1835 Thomas Cole declared every landscape without water defective. The American traveler, dependent upon his garden-park imagery, experienced spatial disorientation—anxiety, shock, terror—triggered by the emptiness of arid America. In arid America the usual markers—trees and water and green colors—were missing, leading to a sense of total loss. There was nothing "artistic" to be seen—no pastures, plowed fields, or picturesque villages; no alpine lakes reflecting snow-capped mountains. It was difficult to use the old visual habits and sense of perspective here; the desert was sensorily austere. Here was a region that flaunted its lack of water and vegetation, its threat to survival. Even advocate Georgia O'Keeffe, the modern desert landscapist, noted that while "a flower touches every heart, a red hill in the Badlands, with the grass gone, doesn't".[59] She used bones and skulls as metaphors for the soilless, treeless land; desert takes us down to irreducible basics.

By 1928 the painter John Marin wrote more positively: "Seems to me the true artist must perforce go from time to time to the elemental big forms—Sky, Sea, Mountain, Plain—to sort of nature himself up, to recharge the battery. For these big forms have everything."[60] In 1965 Andrew Wyeth wrote, "I pre-fer . . . when you feel the bone structure in the landscape—the loneliness of it."[61] In Canyonlands prosaic ground cover is re-moved as nowhere else, and one is allowed to see the real stuff of eternal America. Life is intensely acute when it is experi-enced without any mediation or gradual awareness. It is as if we suddenly know too much, as in a religious revelation. The desert, lacking mediating cover, is a place of purgation, chastity, and discipline.

American antiquers did gradually enter America's desert re-gions, and their numbers have swelled today to unmanageable proportions. By the 1870s and 1880s folks who had been con-demned as tubercular, dropsical, and scrofulous, troubled by weak hearts, disabled lungs, and worn-out nerves, were being miraculously revitalized by the cure of arid and uninhabited regions. My wife Barbara's great-uncle, given a year to live at age twenty because of his wheezy lungs, moved from the moldy Midwest to the small town of Tucson, Arizona; he never re-turned and lived to be over a hundred years old. Not everyone was cured, however. One Easterner who ventured into the dry country complained that he came back an unintended twen-ty pounds lighter, "skin like a chip, juices dried in me, nerves tense, and brain on fire."[62]

The desert not only affected miraculous healings of the body but also brought spiritual renewal. Had not Jesus and Mohammed undergone their crucial cleansings in desert wil-derness, not to mention Moses's theophany on Mount Sinai? Had not the admirable ascetic discipline of early monasticism gotten its start among desert hermits in Upper Egypt? Desert became the environment of revelation, the archetype of sacred space, symbolizing purity and timelessness. The Western desert became America's Holy Land, Canyonlands a metaphor for the eternal and the infinite. Human time is overwhelmed by pri-mordial time, the world of unbelievably slow rhythms now open

18. Pioneering geologist Clarence Dutton of the U.S. Geological Survey found the Grand Canyon sublime, naming many of its major features after powerful gods from the world's great religions. Neither Dutton nor his mentor, John Wesley Powell, saw any difficulty in conflating science with the sacred drama of the canyon. Both were eloquent writers who preached to a large, receptive public. Historian Donald Worster writes: "Dutton's names are meant to reduce the incomprehensible to the manageable. Yet he acknowledges that names cannot really capture the overwhelming power of nature in this place—a nature whose antiquity far exceeds the most ancient human civilizations and whose power can make even industrial America catch its breath." Dutton named the formations, Powell ran the canyon's river, and the polymath William Henry Holmes created the line drawings published in both Dutton's atlas and Powell's tale of his canyon adventure. The combination provided the American public with a vivid virtual reality of a true Western wonder.

to view. This confirmed eternity as the foundation for America's abbreviated national history. This discovery was ontological: "So that's what the world is really like." A bold, empty, silent terrain stood at the center of a theophany. Canyonlands landforms have been from the earliest days interpreted as monumental architecture, akin to the pyramids of Egypt, Stonehenge, or the figures of Easter Island. John Wesley Powell, in his dramatic wooden-boat exploration of the Grand Canyon, named canyon formations after temples: Isis, Buddha, Brahma, Zoroaster, Walhalla, and Deva. For Americans, lacking a classical past but hankering for one, the landforms became America's ancient ruins out of a legendary past. Arizona's aptly named Monument Valley, for example, provides the backdrop for innumerable Western movies.

Environmental historian Steve Pyne, in *How the Canyon Became Grand*, describes the ways in which Americans gave the Grand Canyon its continuous aura of wonder. He reports a sequence of canyon visions tied to historical circumstances. At first, beginning with Spanish explorers and missionaries, the canyon was treated like an immense geographic freak that stood in the way of empire and deserved to be avoided, forgotten, and dismissed. Early American explorations, led by Clarence Edward Dutton, Joseph Christmas Ives, and especially John Wesley Powell, looked upon the canyon for inspiration, as well as for topographical information. The canyon's bizarre depth and length became a natural treasure instead of a valueless impediment. Pyne evokes "a cultural canyon, the Grand Canyon as a place with meaning, shaped by ideas, words, images, and experiences instead of faults, rivers, and mass."[63] Still, few paid attention to Native American creation stories set in the Grand Canyon area or to its continuing sacredness within Native American culture.

The Grand Canyon, in U.S. culture, was not so much revealed as invented, a combination of explorers' exclamations about this sublime place and the paintings, etchings, and lithographs constructed by the romantic landscape artist Thomas Moran, who accompanied Dutton's expedition as its pictorial reporter. Moran's monumental canyon landscape depicted a high spiritual truth in a physical reality: "no one would mistake a Moran Canyon for anything but the Grand Canyon."[64] On his vast canvases Moran offered Americans an idealized canyon that was closer to dramatic art than to the actual place, "an opera of almost Wagnerian excess." And soon the mightiest nineteenth-century technology—the railroad—brought excited tourists to the canyon. The canyon became premier among America's many natural marvels—grander than Virginia's Natural Bridge or New York's Niagara Falls, even more monumental than Yellowstone or Yosemite, far more powerful than anything Europe could offer—and demonstrated the superiority of America's manifest destiny.

At the Grand Canyon one had a sense of being "present at the creation," at the end of time, and during all ages in between. The dimension of time, in other words, penetrated the dimension of space. Fleeting human events, even the spans of civilizations, were but a temporary efflorescence compared to the tempo of the rocks. The Grand Canyon, then, became not a metaphor but the most immediate example of eternity and infinitude.

Eco-Antiquing

By the 1970s Americans were seeking out sites that gave life to a new environmental awareness of natural diversity and ecosystem complexity. A century earlier an explosion of scientific knowledge about the West had come from the great Western land surveys led by Clarence King, George Montague Wheeler, Ferdinand Vandiveer Hayden, Nathaniel P. Langford, Clarence E. Dutton, and John Wesley Powell.[65] These had changed

American attitudes toward the value of wilderness. Yet commitment to preservation remained difficult for Americans, since, in David Lowenthal's words, "action became so strong a component of the American character that landscapes were often hardly seen at all; they were only acted on."[66]

Genteel resorts did not foster proper intimacy with the wild. When Ralph Waldo Emerson visited John Muir at Yosemite, he barely ventured beyond the lodge's porch, so Muir was thrilled with Theodore Roosevelt's demand to hit the backcountry. Such "roughing it" was not a reenactment of the pioneer camp or a recapitulation of the wilderness frontier experience, but instead elemental contact with the reality of nature.[67] Roosevelt's enthusiasm for "the free, open, pleasantest, and healthiest life in America" was infectious. At times, however, such enthusiasm could also become merely lighthearted and trivial antiquing. A company of city folk, finding and preparing food, washing clothes in the river, building fires, and going barefoot, gaily related, "We played we were French peasants, and what fun it was!"[68] Even Muir's early Sierra Club outings were more like catered picnics than serious ventures. More primitive camping and backpacking did not take hold under the 1960s.

The eco-antiquing of the 1970s went beyond tourism's usual quests for monumental scenery, primitive outdoor recreation, and watchable wildlife. A good deal of 1970s environmental rhetoric focused on a now-classic question: "Do trees have rights?" Tourists, listening to the park naturalist, felt the urge to become activists. Their environmental aesthetic might focus on visual values intrinsic to a working ecosystem; the related affirmation might be, "A healthy environment radiates beauty." Did places still exist where nature's organic churning was largely left to itself? The WildEarth movement went so far as to urge "back to the Pleistocene"—to the way things were at the onset of human emergence.

Ecotourism's vision can be defined, in part, by what it is not.

We certainly can exclude *recreation*, which can mean noisy jet skiing on Lake Mead or beer guzzling on a houseboat on Lake Powell. Park officials at Zion National Park in southwest Utah, to their credit, eliminated as inappropriate a swimming pool in the 1970s and tennis courts in the 1980s. Ecotourism can also exclude *adventure*, such as rowing a raft down the Colorado River rapids through the Grand Canyon National Park, mountain biking over slickrock at Arches, or climbing El Capitan at Yosemite. Ecotourism also excludes, perhaps with greater difficulty, the traditional *wilderness appreciation*—seeing scenic wonders—that did so much to establish national parks in the nineteenth and twentieth centuries.[69] Sublime mountain or forest views are wonderful, as are most landscapes free of human artifacts, but they may have little to do with a self-sustaining natural ecosystem. John Muir, for example, complained bitterly when Washington's Mount Rainier National Park was created in 1899, since it specifically excluded the surrounding forests because of their commercial value. He wrote, "The icy dome needs none of man's care, but unless the reserve is guarded the flower bloom will soon be killed, and nothing of the forests will be left but black stump monuments."[70]

Unique natural places can offer good reasons for national parks, but they are not to be confused with the completeness of ecosystems. The Yellowstone ecosystem is larger than the borders of the national park; ecosystem boundaries follow watersheds, animal habitats, and climate vegetation zones, not the straight lines and ninety-degree land-survey corners of park borders.[71] Some national park ecosystems may be beyond repair—Yosemite, the Great Smokies, Shenandoah. A new virtual reality—ecosystem science—has introduced a new lens through which to peer at that strange outside world.

Antique America marked the beginning of attention to the natural world as a place that deserved protection, instead of the far more typically American exploitation. Americans overlaid Nature with multiple strata of imagination. The question of whether this imaginative tradition can be supplanted by a clearer vision of the reality of Nature is one of the primary themes of the last chapter of this book.

3

Human Kodaks in the Future Perfect
Virtual America Embodied in World's Fairs

Americans, crowding into their world's fairs, could not say enough about the thrill. The fairs felicitously combined America's engineering prowess, the material success of consumerism, and the nation's Manifest Destiny. Indeed, to the burgeoning flow of immigrants all of America was a fair. In fair historian Robert W. Rydell's words: "Far from simply reflecting American culture, the expositions were intended to shape that culture. They left an enduring vision of empire."[1] As Americans strolled through fair exhibits, they felt superior to curious European villagers and pitied "primitive" African dance groups. "Exotic" Asian handicrafts seemed only old-fashioned. America's fairs exhibited a technological sublime, a heightened level of accomplishment, the latest technologies promising pathways to utopia. The fairs, as fantasy worlds incarnate, gave Americans an explanatory blueprint of the future.[2]

"Progress" was the official theme of the first modern world's fair, the famed 1851 Great Exhibition in London. This exhibition, which set the standard for future fairs, centered on the magnificent glass and iron Crystal Palace. According to the fair's sponsor, Prince Albert, it displayed the world's greatest industrial achievements. Visitors were stunned by the deafening noise of labor-saving locomotives, marine engines, hydraulic presses, power looms, and printing presses, all powered by steam. A model of James Watt's original 1785 steam engine—one cylinder, forty horsepower—was placed alongside a modern marine engine—four cylinders, seven hundred horsepower.[3] Symbolically, a single huge block of coal, weighing twenty-four tons, sat majestically in the hall, like a gorilla from First Nature, to be gawked at from the safety of Second Nature.

19. Americans, and everyone else, were astounded by the seemingly impossible brilliance of electric lighting as it filled the White City at night: one hundred thousand incandescent bulbs, five thousand arc lights, and twenty thousand glow lights that presented "a fairy scene of inexpressionable splendor." Nothing remotely similar had been seen before. *Harper's Bazaar* reported on "the wonderous enchantment of the night illumination . . . one fails for want of proper words" (September 9, 1893). George Westinghouse had battled Thomas A. Edison for the lighting contract for the fair, the former advocating the newfangled alternating current while the latter stuck with his proven direct current. Westinghouse won, and the history of electric lighting was revolutionized. He worked the entire fair from a single central switchboard, in itself a step into the future.

THE CORLISS ENGINE IN MACHINERY HALL.

20. A great block of coal stood in London's Crystal Palace as the centerpiece of its 1851 Great Exhibition. This world's fair was the model for America's 1876 Centennial Exhibition in Philadelphia, whose centerpiece was George Henry Corliss's enormous steam engine, seen as a mark of the tremendous technological progress that had been made in the previous twenty-five years. Thirty feet high, it weighed seven hundred tons and produced fourteen hundred horsepower. This single engine powered the machinery of the fair through a central drive shaft that moved secondary belts and pulleys. Corliss's great engine told Americans of their new industrial power, which guaranteed a superior standard of living and quality of life—a physical representation of America's superiority to the rest of the world.

To anyone strolling through the rows of machines, it seemed as if, at long last, history had become the story of humanity's inevitable advance. Nature had disappeared behind the machines.

Americans put together their own Great Exhibition, the Centennial Exhibition in Philadelphia. This 1876 exhibition hardly celebrated 1776; instead its structures and exhibits depicted "a working model of an America Mecca." Congressman Daniel J. Morrell concluded that America's epic battle with nature had ended in permanent conquest. Instead of a block of coal, the Centennial Exhibition's centerpiece was George Corliss's huge steam engine, the largest in the world. President U. S. Grant, joined by the visiting emperor of Brazil, inaugurated the fair by turning this engine's handles, building up steam pressure. The engine's great arms—thirty feet high, weighing seven hundred tons, and producing fourteen hundred horsepower—slowly accelerated to produce power for all the exhibits through drive shafts, belts, and pulleys. The president and emperor stood back in awe. The Corliss engine embodied the American Dream. Public opinion compared it to a great heart whose steady pulsations guaranteed a seemingly impossible future—a steady, uninterrupted flow of energy to power American ambitions. The engine stood before the Centennial Exhibition audience as evidence of the triumph of material progress through technological innovation and industrial enterprise since the American Revolution.[4] Had not Thomas Jefferson first introduced the phrase "the Pursuit of Happiness" into the Declaration of Independence, meaning material abundance? In a rhetorical stretch Congressman Morrell claimed that "it affects the imagination as realizing the fabled powers of genii and afrit in Arabian tales, and like them it is subject to subtle control."[5]

Even the crusty Henry Adams, so alienated from American society that he had escaped to England, admitted that world's fairs were their own religion. Philadelphia's Centennial Exhibition had opened with a concert of Richard Wagner's operatic music, based on great Germanic and Norse mythologies, but no one doubted that the soaring orchestral themes pointed to America's Manifest Destiny, its emergent global hegemony and eventual worldwide triumphalism. Rydell writes that the "fairs resembled religious celebrations in their emphasis on symbols and ritualistic behavior."[6] The Reverend D. Otis Kellogg reported that the Centennial Exhibition's Main Hall's power to evoke the presence of the divine could be positively compared to that of a great European cathedral. People further looked to rewards from the dynamo, which had more of an "aura" than did the statue of the Virgin Mary at the heart of Chartres cathedral.[7] The *Philadelphia Press* urged visitors, "Let us, therefore, today bare our heads and take off our sandals, for we tread on holy ground."[8] Writing about the 1893 World Columbian Exposition in Chicago, *Chicago Tribune* columnist George Alfred Townsend declared its displays "the confessions of faith of a new dispensation."[9] America was God's chosen nation. Most contemporaries also reported a "moral influence" emergent from any visit to the 1893 exhibition. Visitors expected to leave the fair refreshed in their personal ethic of duty and benevolence. Rydell too finds a moral compass attached to the fair, "a fixed coordinate about what social, political and economic choices could be judged as right or wrong, good or bad, in the past, present, and future."[10]

Fairs and exhibitions since the Middle Ages had always celebrated an idealized fantasy world that surfaced, albeit fitfully, in the meaner world of hard labor, dreary lives, and early death. Such fairs provided secular versions of the heavenly delights promised in the great cathedrals—the divine presence in the sacraments, and biblical stories brilliantly portrayed in stained-glass windows. By the time of the self-indulgent Renaissance, celebrations of a fleeting, earthly paradise involved great pageants in brilliant city squares. The clothing of the upper classes

—the men in multicolored tights and the women with over-flowing bosoms—attempted to bring a dreamworld into real life, as did the stupendous and detailed architecture of the Renaissance, from Saint Peter's in Rome, to Florence's Duomo, to the Oriental opulence of Venice's Saint Mark's cathedral. The modern world's fairs, then, had many models upon which to build: religious festivals, circuses, zoological gardens, minstrel shows, and later sanitary fairs, industrial exhibits, and even Wild West shows.

America's world's fairs sprang up like dandelions. Each had a similar life span: San Francisco in 1894; Atlanta in 1895; Nashville in 1897; Omaha in 1898; Buffalo in 1901; Saint Louis in 1904; Portland, Oregon, in 1905; Seattle in 1909; San Francisco again and San Diego in 1915; Philadelphia in 1926; Chicago in 1933; San Diego again in 1935; New York in 1939; San Francisco a third time in 1939; Seattle again in 1962; New York redux in 1964; San Antonio in 1968; Spokane in 1974; Knoxville in 1982; and New Orleans in 1984. By the 1980s and 1990s the dreamworld had gained permanence with the unveiling of Florida's EPCOT and Minnesota's Mall of America.

Nothing else, however, would compare to the extravagant dreamworld offered by the world's fairs—"symbolic universes"—until the appearance of today's cyberspace. The fairs allowed visitors to stroll through a virtual reality that had materialized before their eyes. The buildings, displays, fountains, lighting, and live shows incarnated utopian dreams-come-true of America's material and national progress. Philadelphia's Horticultural Hall was "as entrancing as a poet's dream," "an

21A and 21B. The cartoonist assumed, even in 1980, that Americans remembered that the classic White City of the 1893 Chicago Columbian Exposition was "heavenly." The second cartoon pokes fun at Disneyworld's attempt to portray a futuristic sanitized America, never quite leaving behind the gritty reality of a local bar with its beery inhabitants. Stevenson suggests that in the contest, the rickety old bar wins.

"Oh, goody! It's just like the Columbian Exposition of 1893."

Arabian Nights' sort of gorgeousness."[11] The brilliant electric lighting of Chicago's 1893 World Columbian Exposition flooded its White City. The *Chicago Tribune* described the neoclassical metropolis as "a little ideal world, a realization of Utopia . . . [a] time when the earth should be as pure, as beautiful, and as joyous as the White City itself." It added, "The impulse which this Phantom City will give to American culture cannot be overestimated."[12] When President William McKinley opened the Pan-American Exposition in Buffalo in 1901, he could have been speaking of today's virtual reality on the computer screen: "Expositions are the timekeepers of progress. They record the world's advancement. They stimulate the energy, enterprise, and intellect of the people and quicken human genius. They go into the home. They open mighty storehouses of information to the student."[13] (A few minutes later McKinley was fatally shot by Leon Czolgosz, a reputed anarchist.)

Indeed, one can argue that America's historic experience with the delights of world's fairs helped open the door to today's virtual reality. Like today's simulations the fairs were mythopoeia; they had the capacity to generate continuously rich and complex visions of "what might really be." Historian Dustin Kidd compares the guiding principles of the ordered space displayed at fairs with the future structures of cyberspace, the Web, and the Internet: "The machine was not only the icon for the machine-age, but also its ordering principle." Cyberspace provides an ordering principle "in which information is ordered and disseminated." He adds: "Just as the machine changed the relationship between man and his reality [as demonstrated at the fair], so today the web is changing our relationship to the world around us. . . . We make it [the world] manageable by collapsing the time and space of information into the ordering space of the World Wide Web."[14]

Like someone entering today's Internet or a simulation, the fair's visitor, according to literary critic Norman Hollander, no longer felt any distinction between "in here" and "out there": "Indeed, there is none; he became 'absorbed.'"[15] The self was blurred into the simulation. Visitors to Chicago's Columbian Exposition were urged to become a tabula rasa upon which the dreamworld would be imposed. In this light visitor exclamations—"mental intoxication," "fairy land," "sacred events"—take on a darker meaning. Mariana G. van Rensselaer, art critic for the popular *Century Magazine*, instructed readers to enter the gates "wholly consciousless—not like a painstaking draftsman, but like a human kodak, caring only for as many pleasing impressions as possible, not for analyzing their worth."[16] One Portland sightseer in 1909 wrote that the fair touched him like an electric shock that "makes my fingers 'all pins and needles.'" It carried the jolt of a mint julep: he felt "mental intoxication." His life of "lack-luster black and white" was transformed into a "spread in God's colors," as if he had shifted from Dorothy's Kansas to the Land of Oz. One children's book author wrote promotional copy for the 1915 San Francisco fair that could be used to advertise today's game simulations: "There was the beautiful tower of Jewels, smiling and twinkling its shiny eyes at us, and saying, 'Come in, children, come in, and walk under my beautiful blue arches, and through my magic courts, and my sheltered gardens, and be happy, and love each other and all the children of the world. Peace I offer you, and Plenty, and Harmony, and Beauty. Here you are safe, and here you are welcome."[17]

If nothing else, world's fairs became fixated on their own hubris. External reality became irrelevant. The fairs not only reported Manifest Destiny; they sped it on its way. Omaha in 1898 claimed to be at the edge of the frontier that stood as the climax of America history—civilization, purified by its Westering trials, moved into the final void. Exposition president Gurdon Wattles announced, "The Great American Desert is no more." Nature's worst challenge, the arid, windblown Great Plains, had become

Arcadia. The fair was a "beacon for all the world" to emulate.[18] At Saint Louis in 1904 the Louisiana Purchase Exposition became "The Coronation of Civilization." Fair official F. J. V. Skiff felt called to proclaim, "The scene which stretches before us to-day is fairer than [that] upon which Christian gazed from Delectable Mountain."[19] At Portland in 1909 the *Oregon Daily Journal* announced, "Where once was wilderness, now blooms paradise . . . a mighty monument to American energy and heroism . . . full-grown Eden." For the awestruck public, the *San Francisco Examiner* concluded, "the fairy tales of today are a thousand times more splendid than the fairy tales we read about in the storybooks."[20]

In an era of strikes, economic setbacks, and political unrest, the 1893 Chicago Columbian Exposition offered an escape into an alternative world of magisterial hope.[21] The 1876 Philadelphia fair had already told Americans how mass production and merchandising would make them the world's best consumers. In 1893 Americans were told that the approaching century would inevitably become, through science and technology, the "American Century." The leading newsman of the day, Richard Harding Davis, called the 1893 fair "the greatest event in the history of the country since the Civil War." The Columbian Exposition became a new dispensation. An official of the 1915 San Francisco fair, Charles F. Lummis, in a letter to David C. Collier, real-estate investor and railroad magnate, declared: "You have mixed Science with Business so splendidly."[22] World's fairs offered showcases of invention and consumerism, mythic dreamworlds that foretold a utopian future.

America's Manifest Destiny became vividly real in the looming full-scale replica of the battleship *Illinois*, docked at the fair's Naval Pier. The grand buildings and heroic sculptures of Chicago's Columbian Exposition openly imitated Imperial Rome, with gold-ribbed domes, rows of columns adorned with banners, and wide imposing staircases. The Court of Honor rivaled Rome's Forum. Architectural critics scoffed, but the public saw symbols of American greatness. The fairgrounds were strikingly similar to Thomas Cole's 1836 landscape *Consummation of Empire*, part of his provocative series *The Course of Empire*, ominously followed by *Destruction and Desolation*. Others were reminded of the formal public buildings in the utopian city of Edward Bellamy's best-selling novel *Looking Backward*. The orator John J. Ingalls invoked the image of Progress: "The most obtuse observer cannot fail to perceive that the path of humanity has been upward . . . that man has advanced further and more rapidly in the last fifty hears, than in the previous fifty centuries. . . . We are living in the best age of history and the most favored portion of the globe. We stand on the summit of time."[23]

The new Electric Age now made Philadelphia's 1876 fair seem almost medieval. Chicago historian Donald L. Miller adds that among Chicago's 1893 displays "there were electric kitchens and calculating machines, electric brushes for relieving headaches, electric incubators for hatching chickens, electric chairs for 'humane' executions."[24] Visitors also marveled at technological revolutions based on electricity: Elisha Gray's telautograph, allowing crude faxing via the telegraph; Thomas A. Edison's Kinetoscope, which showed primitive movies. The White City anticipated both modern shopping malls and EPCOT Center. It entertained the public with the Ferris wheel and drew them into gawking and shopping along the shabbier stalls of the neighboring midway.[25] Lyman Gage, a Chicago bank president, told the planning committee that the fairgoer "will see beautiful buildings radiate with color and flashing the sunlight from their gilded pinnacles and domes. . . . And beyond all, [fairgoers] will behold the boundless waters of Lake Michigan, linking the beautiful with the sublime, the present with the past, the finite with the infinite."[26]

The difference was lighting—as Pittsburgh's George Westinghouse cannily realized when he gained the contract to illuminate the White City. An outlandish alternative world—night into day—was becoming reality. Chicago's White City, brilliantly lit by thousands of electric bulbs run from Westinghouse's central switchboard, was the future incarnate. The novelist Robert Herrick wrote in wonder: "The long lines of white buildings were ablaze with countless lights. . . . The people who could dream this vision and make it real, those people . . . would press on to greater victories than this triumph of beauty—victories greater than the world had yet witnessed." Chicagoan Hilda Satt recalled that when the "millions of lights were suddenly flashed on, all at one time . . . [it] was like getting a sudden vision of Heaven."[27] Canals, lagoons, and basins reflected the electric brilliance.

Chicago offered the first elevated electric railroad in 1883, so popular that its lines—including the "Loop"—remain in use today. The nation's interurban electric railroad system allowed travel that could carry an intrepid traveler from Portland, Maine, to Sheboygan, Wisconsin, with many connections, through major Eastern and Midwestern cities. Americans talked with each other over extensive Bell Telephone networks. The construction of skyscrapers depended not only on steel beams but also on Otis's electric passenger elevator, first installed in 1892. For its shoppers Wanamaker's innovative Philadelphia department store had already installed carbon-electric lamps in 1878.

The new magic of electricity was also coming home. Americans woke up to a deliriously new day—a wonderland that made ordinary citizens feel like sumptuous emperors. Lighting and heat day and night; speedy, quiet, and easy travel; multifold ways to enjoy music; less hard physical labor; electric appliances —the list seemed endless. Americans in all walks of life rushed to remove from their households the old world of expensive candles, sooty lanterns, and explosive gas lamps. Electricity provided energy "on tap" when required (although electricity would not reach some rural areas until the 1930s and 1940s).

After 1880 electric power surged on full stream, compared to the mere trickle of the telegraph and telephone. Domestic electric consumption rose in the 1890s from virtually zero to nine billion kilowatt hours, which wrought dramatic effects—household lighting, electric irons, refrigerators, electric bread toasters. Also emerging was a renewed sense of national well-being. The United States now dominated the all-powerful electric mass media of commercial radio, talking movies, electric signs, and later television and computers. The impact of electricity upon a transformed America cannot be overemphasized.

Metropolitan America: *Mythical, Real, Simulated*

One of today's simulations—*SimCity*—involves the building and development of an idealized city. Games philosopher Ted Friedman describes *SimCity* as resting on "the empiricist, technophilic fantasy."[28] The Maxis catalog states that *SimCity* "dares you to design and build the city of your dreams. . . .Depending on your choices and design skills, Simulated Citizens (Sims) will move in and build homes, hospitals, churches, stores and factories, or move out in search of a better life elsewhere."[29] The simulated reality is built from the ground up, from an undeveloped patch of First Nature.

America's fairs invariably depicted a metropolitan utopia, never a rural Arcadia. The reordering of America's geography from its agricultural and rural appearance had formed a new order. First Nature was judged of little value until it had been absorbed into Second Nature. The geography of the countryside *between* the cities lost its "presence," becoming a mere resource to be exploited by the new urban imperialism. Cities themselves epitomized a *built environment*, overlaid on the natural environment, made up of streets, gutters, and curbs; sidewalks and street lights; and the walls and windows of concrete, steel, brick, and wooden buildings. Below the surface was the "underground city" (Lewis Mumford's term): a hydraulic, pneumatic, and electrical maze of subway and service tunnels, tracks, platforms, storage rooms, and shops, as well as water mains, sewers, gas mains, and the ganglia of power lines. Thus, the natural terrain of the city receded into the background. Overall, Americans began to live and work and play in fabricated networks more than they inhabited the natural world.

In *SimCity* you are Mayor or City Planner; in a similar game, *Civilization*, you are Chief or Warlord, Prince or King or Emperor; in *Populus* you are God. In all cases your powers are godlike: you build power plants; construct industrial, commercial, and residential areas; lay down roads, mass transit, and power lines; and provide services like police and fire departments, airports and seaports, and even sports stadiums. You collect taxes and devise economic strategies, with the goal of building a large population and winning high approval ratings from your Sims citizens. *SimCity* is self-regulating.

One game company, Activision, in planning *True Crime: New York City*, used six location scouts holding detailed maps to walk the borough of Manhattan, photographing practically every intersection and major landmark. Using eleven thousand images, they sought to accurately re-create an existing place: "We tried to make it a real living, breathing city."[30] While the simulation includes little details like construction scaffolding and trash cans stuffed with trash, its most unrealistic aspect is the absence of traffic, access to taxies, and street life. The game's producer, Simon Ebejer, admits that the game reflects out-of-towners' stereotypes: "We made the city grittier and dirtier than it is in real life because that's the common perception of what New York City is like." Writer Daniel Radosh observes that "the fun of a simulation like *True Crime: New York City* is not that it re-creates the city perfectly but that it re-creates the city just well enough to allow a player to have adventures that are forbidden in real life."[31] Or impossible in real life, reminiscent of an eve-

22. This image is from the 2006 University Expansion Package for *Sims* 2, meant to appeal to young adults focused on a college lifestyle and personal career development. It remains based on the original *SimCity* (1989), which has morphed into *SimEarth, SimFarm, SimTown,* and *SimAnt,* as well as the grander, empire-building *Civilization,* the business-model-focused *Capitalism,* and the biocentric *Spore,* where both humans and the natural world interactively participate with the player. In all these versions an infrastructure inhabited by active (artificial) people, with its own responsive dynamics, is central to the simulation. There are also versions based on actual historical events, such as the 1906 San Francisco earthquake and the 1944 bombing of Hamburg during World War II, and versions based on possible disasters in existing cities, like a 2010 nuclear accident in Boston or the 2047 flooding of Rio de Janeiro. One unauthorized version, *SimSim,* allows players to design simulators, an example of "deep simulation." Players describe the *Sim* experience as far more affective than the one-sided exchange of reading a book, the disengagement of word processing, or the limited feedback of surfing the Web. The simulation is deliberately designed to induce a creative positive response, to "build the city of your dreams," to "search for a better life."

ning stroll through the brightly lit White City, or an adventure at the scruffy Midway of Chicago's 1893 World Exhibition.

From Utopia to Dystopia

But all such utopianism—all virtual reality dreamworlds—falters. As Chekhov has Baron Tuzenbach say in his play *Three Sisters*: "Well, maybe we'll fly in balloons, the cut of jackets will be different, we'll have discovered a sixth sense, maybe even developed it—I don't know. But life will be the same—difficult, full of unknowns, and happy. In a thousand years, just like today, people will sigh and say, oh, how hard it is to be alive. They'll still be scared of death and won't want to die."[32]

Chicago's 1933 "Century of Progress" unwittingly portrayed an Orwellian nightmare world. This fair deliberately compared itself to Philadelphia's 1876 fair. Instead of featuring levers that were turned to start a Corliss engine, the opening ceremony of June 1, 1933, unveiled a new world by the flipping of a switch that activated a photoelectric cell that captured the light from the star Arcturus. This wave/particle of light had reached the earth after forty years of travel through space, just in time to power the fair. Americans were convinced they had mastered space and time in all their forms.

The fair's slogan arched over the main entrance: "Science Finds—Industry Applies—Man Conforms." The fair instructed Americans about human conformity to the new "ordered space" of the Machine Age. The virtual realities behind world's fairs, which praised the liberation offered by technological and industrial prowess, also depicted a technology-forced social life. Chicago's 1893 Colombian Exposition, in Robert W. Rydell's words, was above all "an exercise in educating the nation on the concept of progress as a *willed* national activity toward a determined, utopian goal. . . . The pathway to the future could be constructed only out of fibers of human will rightly informed."[33]

The "People of Plenty" would exist only if they rigorously followed the narrow path of industrial production.

Nightmares of totalitarianism and the control of passive populations loomed over the 1939 "World of Tomorrow," built on a reconstituted trash heap in the Flushing Meadows area of Queens, one of New York City's five boroughs. The apparent failure of democracy during the Depression and the specters of flashy German Fascism and determined Russian Communism had eroded the American dreamworld. How to shift from the Depression's tailspin into dystopia and rise into a reconstituted American Utopia? The 1939 fair involved the most deliberate plan yet conceived to turn a specific virtual world into a physical reality. According to a University of Virginia Internet project exploring the 1930s, "the fair had attempted to transform itself into the literal world of the future by providing a very clear vision of the chaos of the past and the purity and peace of the socially-planned future." The commentator John Crowley wrote in 1939: "Actually, Tomorrow scared me a little. Could I grasp the immense plan expressed in occult symbols all over the fair? Would I be up to tomorrow? It seemed so urgent that Tomorrow be dragged out of the Future where it lay, peacefully unborn. But why was it so urgent? Why?"[34] The industrial pavilions instructed the visitor, "You *will* be excited"—not so far distant from the Soviet demand of its citizens, "You *will* be happy!" At the same time it is no coincidence that in the same year, 1939, the classic film *The Wizard of Oz* portrayed the dreamy Emerald City. Recent commentators on the 1939 fair have noted that it anticipated Disney's Magic Kingdom, large shopping malls, and EPCOT Center.

Even the most promising utopian dreamworld, however, collapsed with World War II. More than sixty countries had been represented at the 1939 Lagoon of Nations, including the Soviet Union, but not Nazi Germany. Soon, however, the Lagoon's pavilions for Czechoslovakia, Poland, and Lithuania were the only remaining representations of their former identities. By the fair's reopening in 1940 France had fallen, and Britain seemed to be the next likely victim. The Soviet pavilion was razed, replaced by the "American Common." The "World of Tomorrow" suddenly seemed irrelevant; the fair closed a financial failure, and its operators filed for bankruptcy. The virtual reality offered by the fair could not compete with actual violence. With later fears of the Cold War, and disillusionment about the capabilities of new inventions, post–World War II fairs were increasingly mere shadows of 1893, 1933, and 1939.

Disney's Perpetual Fantasy World

When California's Disneyland opened in July 1955, America's favorite cartoon filmmaker, Walt Disney, said: "I don't want the public to see the real world they live in while they're in the park. I want them to feel they are in another world."[35] Disneyland's 244 acres were walled in by a landscaped but enormous earth embankment to keep the real world out. Disney, like God, saw his creation and said it was good. It was clean, efficient, safe, and controlled.

On February 2, 1967, Disney, in a filmed announcement, told the public that he would now build the world's first glass-domed city—a fifty-acre, air-conditioned "city of tomorrow," itself at the heart of a one-thousand-acre industrial park, a completely closed environment with a minimum of traffic: "The pedestrian will be the king." Cars and trucks would travel underground: "I'm not against the automobile, but I just feel that the automobile has moved into communities too much."[36] High-speed monorails would transport workers from their hub offices to three outlying areas featuring high-density apartments; freestanding futuristic homes; and a green belt including churches, playgrounds, and schools. Disney, remembering the fleeting model city of New York's 1939 world's fair, this time offered permanence. Utilities would include a vacuum trash-disposal system and an underground utility tunnel. A central computer system would control everything from streetlights to hotel reservations. Disney's project was designated the "Experimental Prototype Community of Tomorrow," or EPCOT.[37] Disney had already covertly purchased forty-three square miles in central Florida—an area twice the size of Manhattan, 150 times the size of California's Disneyland. Disney did not spell out what human living entailed, but he was convinced that he would continue to find the heartbeat of ordinary Americans, just as he had done with Mickey and Donald.

Walt Disney, however, had actually already died—on December 15, 1966, six weeks before his February announcement. It was fitting, perhaps, that his utopia was presented on film, his primary medium, accompanied by his disembodied voiceover narration, "EPCOT will always be a showcase to the world for the ingenuity and imagination of American free enterprise." In print he added: "In EPCOT there will be no slum areas because we won't let them develop. There will be no landowners and therefore no voting control. People will rent houses instead of buying them, and at modest rentals. There will be no retirees. Everyone must be employed. One of our requirements is that people who live in EPCOT must keep it alive," servants to the simulation.[38]

Disney, in his planning, shrewdly kept to the mainstream of the 1930s fairs, with their emphasis on conformity and obedience. One critic notes: "Disney intended to regulate the lives of the city's residents almost as thoroughly as the climate in the dome. Representative local government was ruled out. No residents were to be permanent. Pets would be forbidden, dress codes would be enforced, and residents would be expelled for unbecoming conduct ranging from drunkenness to unmarried cohabitation."[39] Disney was insistent: "EPCOT . . . will never cease to be a living *blueprint of the future*, where people actually live a life they can't find anywhere else in the world today."[40] The dreamworld briefly displayed in Chicago in 1893 and 1933, and in New York City in 1939, would now become a permanent model for an ideal future.

Disney sought, and his company achieved, total control, isolated from outside political and social forces. Critic Joshua Wolf writes that "in his own space, in his own way, he wanted to *create not a virtual reality but an alternative reality* that conceded no element of its environment to chance or outside influence. Political insulation was hardly incidental to this vision. It was a central part." Wolf adds, "Disney's civic design may be an acceptable framework for operating a fantasy based theme park, but should it really be, as Walt apparently intended, a blueprint for the public spaces of tomorrow?" Wolf concludes that at EPCOT "the boundaries of the genuine and the illusory are collapsed, defined, and erased all over again."[41] Social critic Ada Louise Huxtable observes of Disney World that "the ostensible purpose of the reproduction, to make one want the original, has been supplanted by the feeling that the original is no longer necessary. The copy is considered just as good and, in some cases, better."[42] Wolf notes, "The Walt Disney empire is more about profits than politics, and this includes relentless tampering with reality."[43]

Continuous Consumerism

Shopping malls as dreamworlds are surely a letdown after the stupendous fabrications of world's fairs and EPCOT. The enterprising Ghermezian brothers, however, believed otherwise, proudly displaying the new world of consumerism in the world's largest shopping mall, remarkably situated in Canada's lonely frontier metropolis Edmonton, Alberta. The appearance of the vast, enclosed West Edmonton Mall evokes that of an alien city dropped from outer space onto flat wheat fields that spread to the horizon, in a region also inhabited by thousands of oil wells and fiercely independent Native American peoples. The mall boasts a size of 48 city blocks; 800 stores and services; 110 places to eat; 26 movie houses, including an IMAX theater; the world's largest indoor water park; a full-scale replica of Columbus's *Santa Maria*; an NHL-size ice rink; submarine rides; a roller coaster; bungee jumping; a miniaturized replica of the Pebble Beach golf course; the requisite Las Vegas–style casino; and boulevards called Europa and Bourbon Street. Its 23,500 employees especially cater to tourists, arriving on packaged visits to this self-proclaimed "Eighth Wonder of the World." By 2005 this inflated enterprise was matched, or surpassed, by the bizarre fantasy world of the Persian Gulf's Dubai.

Mall visits reinvented tourism, spinning it away from its traditional emphasis on the quest for natural wonders or historic sites. Tourism is no longer only "antiquing"; it is now focused on packaged entertainment and the desire for insulated environments where one can inhabit the future. The Ghermezian brothers next installed a mall dreamscape at the vast Mall of America outside Minneapolis, enclosed to cope with long blizzard-filled winters. It opened in August 1992 and now stretches over 4.2 million square feet containing 520 stores, 86 restaurants, 8 nightclubs, and standard child entertainment centers, employing over 12,000 people. Its Web site asserts that the "Mall of America is one of the most visited destinations in the United States, attracting more visitors annually than Disney World, Graceland, and the Grand Canyon combined."[44] Tourists account for four out of every ten visits. Indeed, there is "no need for heat—Mall of America's guests, along with miles of lights, provide enough warmth to keep the entire complex toasty-warm even during the cool [*sic*] winter season." The mall's promoters also claim that, while the mall generated 34,700 tons of waste in its first five years, it successfully recycles more than half of its annual waste. Again, the fantasy world of the Mall of America is stretched to the breaking point by Dubai.

The University of Virginia's study of world's fairs concludes, "The Fair was the modern advancement of American enlightenment ideas: a democratized promised land, intended to uplift all citizens and serve as a shining example for the rest of the world, promoting the notion that economic prosperity results for all who believe." The seventeenth-century New England Puritan "city set upon a hill" had been secularized. After World War II ended, when American consumerism reenergized itself, utopia reappeared in shopping malls. Now, since the 1991 fall of the Soviet Union, we can combine American Consumerism with American Triumphalism.

Conclusion: *Utopia around the Corner*

The momentum toward a world of material well-being displayed by the fairs has continued to shape American society. "Food, shelter, and clothing for all" seemed to be not merely a feature of America's virtual reality but a goal that had been reached. One economist, David A. Wells, looked back over the previous thirty years from the vantage point of 1889: "Man in general has attained to such a greater control over the forces of Nature, that he has been able to do far more work in a given time, produce far more product, measured by quantity in ratio to a given amount of labor, and reduce the effort necessary to insure a comfortable subsistence in a far greater measure [than had ever before been possible]."[45] A publicist for the chemical industry, Edwin P. Slossen, wrote an article in 1920 entitled "Back to Nature? Never! Forward to the Machine!" He announced, "The conquest of nature, not the imitation of nature, is the whole duty of man."[46] Nature had ceased to be life's master. By the middle of the twentieth century Americans were convinced that industry, technology, and the spirit of enterprise were the tools necessary to capture and enjoy the wealth entrapped within nature. The distance gained from the raw forces of nature measured personal success. World's fairs, EPCOT, and shopping malls seemed to demonstrate humanity's victory. Simulations like *SimCity* turned this victory into entertainment that could be inhabited.

Americans came to believe that they had closed in on Utopia. Historically, Western Civilization has always needed dream-worlds to maintain its momentum. Utopianism involves the ineffable and inexpressible spiritual joy described by the Greek Dioysius the Areopagite, Judaism's New Jerusalem, and the Italian Dante's *Paradiso*. Like today's habitation of simulations in cyberspace, this bliss is indeterminate. It can be endlessly created and re-created in the imagination.[47] America's Manifest Destiny and global hegemony were meant to produce a worldwide state of grace. Secular versions of a heavenly virtual reality have appeared in science fiction, world's fairs, and especially a variety of utopian visions. But Utopias like those detailed in Plato's *Republic*, Thomas More's *Utopia*, and Edward Bellamy's *Looking Backward* have dismal records, while Aldous Huxley's *Brave New World* and George Orwell's *1984* are dystopias, and Soviet and fascist utopian dreams cannot be judged any better. If accruing wealth is the primary utopian agenda, then the United States fits the bill. This was recognized in the nineteenth century by Willa Cather in her story "Paul's Case," which describes a man who gives up his whole dreary life for a few days of luxury, similar to a vacation in EPCOT, Las Vegas, or Dubai. As for rural "green" utopias, once devised by William Morris in *News from Nowhere* and H. G. Wells in *The World Set Free*, and still fostered in such works as *Walden Two* and the novels of Ursula Le Guin, they falter on hard labor and the support necessary for growing populations. And technological utopias, so wondrously displayed in world's fairs, and once praised by Herman Kahn and other think-tankers, seem to find completion only in today's American triumphalism and hegemony.

4

Sleepwalking in America

A Brief History

The story thus far: we have discovered that today's cyberspace provides a useful parallel to American history, cloaked in mythic images. Today a virtual reality—an island in cyberspace—is, for many of us, a working alternative home, encompassing search engines on the Internet, e-mails and blogs, simulations and games. Our participation in synthetic cyberspace can, paradoxically, deepen our understanding of our history set in concrete geography. While virtual reality enlivens and enriches the debate over "home place," it nevertheless has made physical location seem even more fleeting and lightweight, relative and indeterminate, leaving us less personally involved than we are in our blogs and simulations. This conundrum will be at the center of the next chapter. Still, cyberphenomena can be added to our toolkit in our quest for genuine habitation of authentic places.

Throughout this book we describe nonhuman geography as First Nature, the built environment as Second Nature, cyberspace as Third Nature, and sense of place—the merger of self and habitation—as Last Nature. In this chapter we explore three facets at the heart of the historic American experience: Engineered America, Consumer America, and Triumphal America. All are primary ingredients of Second Nature. They are also intimately interlocked. And while all have their virtues, while all have done much to define America, they are also flawed, and their limitations inhibit our quest to inhabit authentic place.

Engineered America

Creating Utopia: The Disappearance of First Nature

ENGINEERS: HIGH PRIESTS OF SECOND NATURE

Leonardo-like for his polymath achievements, John August Roebling stood tall in post–Civil War America as a national celebrity, a civil engineer who had fulfilled American dreams of mastering nature. As much as any engineer had a specialty in his day, his was dramatic suspension bridges.[1] His most famous spanned the turbulent East River between the bustling cities of Brooklyn and New York. Previously New Yorkers had been frustrated because only numerous ferryboats and barges, so well described by Walt Whitman, connected the burgeoning cities. Roebling's Brooklyn Bridge, completed in 1883, stretched almost a mile between the bridge approaches and the river. According to one promoter, it took "one grand flying leap from shore to shore over the masts of the ships."[2] The bridge would make up for what nature had failed to provide. As Americans conquered rivers, mountains, deserts, and scorching plains, they began to see nature less as a worthy adversary than as raw material to be modified for their consumption. Tellingly, when England's famous scientist Lord Kelvin visited Niagara Falls and was asked whether hydroelectric power would detract from the site's natural beauty, he retorted: "What has that got to do with

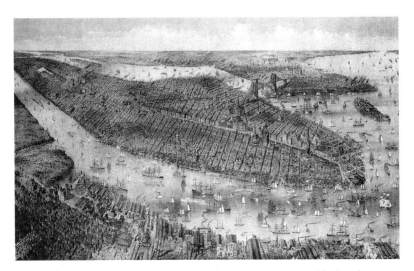

23. Bird's-eye views of America's cities were magical images, considering the height of the perspective—far above a rare balloon ride—and the fact that they came decades before heavier-than-air flight was realized. This popular 1875 Currier & Ives print by Frances Palmer, one of the most prolific Currier & Ives artists, shows an extravagantly overscale bridge that appears to drag New York City closer to its sister city, Brooklyn. The image of the Brooklyn Bridge was so powerful in evoking the new engineered America that Currier & Ives would eventually offer 16 different colored lithographs featuring it, along with 150 different renderings of New York City between 1835 and 1907. As for New Jersey, it was a stepchild, with only ferryboats linking it to the other side of the Hudson. This view, however, shows multiple railroad termini hankering to leap the river.

it? I consider it almost an international crime that so much energy has been allowed to go to waste."[3]

Soon the engineer emerged as the quintessential American hero. He (rarely she) began to receive more esteem than the traditional professional clergyman, doctor, or lawyer. Engineers became an elite class, little different from the high priests who had acted as the managers of the great constructions of ancient civilizations. The engineer's efforts were all in the cause of the transition from European squalor to American paradise.

Roebling personified the larger-than-life image of the engineer: inventive, pragmatic, skilled, self-made, prosperous, and highly ethical. David McCullough, biographer of the Brooklyn Bridge, describes Roebling himself as physically impressive, noting that he "gave the appearance of having been hewn from some substance of greater durability than mortal flesh . . . a man of enormous dignity, plainly enough, full of purpose and iron determination."[4] A fellow engineer wrote, "One of his strongest moral traits was his power of will . . . a certain spirit, tenacity of purpose, and confident reliance upon self . . . an instinctive faith in the resources of his art that no force of circumstance could divert."[5] But Roebling's enormous self-confidence became his downfall when he rejected medical assistance after an on-site construction accident crushed the toes of his right foot. He had his toes amputated without anesthetic, insisted on treating himself, and died an awful death from tetanus a month later, in July 1869.[6] The *Brooklyn Eagle* reflected the sad mood of a grateful nation when it declared, "He who loses his life from injuries received in the pursuit of science or of duty, in acquiring engineering information or carrying out engineering details, is . . . truly and usefully a martyr."[7] Roebling's hubris had left him with feet of clay.

Another engineering hero on this same divine level was Rudolph Hering. Hering, literate and philosophical, had been a premier student of the great German philosopher Friedrick Hegel in Berlin, and he followed Hegel's emphasis on self-realization, the supremacy of rationality, and the importance of fulfilling a great destiny. After training in Dresden in Prussia in 1868, Hering migrated to the United States to begin his career as a surveying assistant for the new Prospect Park in the city of Brooklyn. His surveying work continued at Philadelphia's new Fairmount Park; brought an excursion to set the boundaries of faraway Yellowstone National Park; and finally took him to Philadelphia's Girard Street Bridge, where he was engineer in charge of construction. Hering exemplified the engineer as social hero who practiced civic virtue when he took charge of

Philadelphia's massive reconstruction of its main sewers and bridges, including breaking the hold of corrupt contractors.

Engineering is, according to one encyclopedia, "the art of directing the great sources of materials and power in nature for the use and the convenience of humans." America's geography was thus transformed from the myth of a dark chaos into a vast body of raw materials, both real and symbolic. The encyclopedia definition adds that engineering "is the useful application of natural phenomena." Natural phenomena left alone are obsolete, but with the potential to be modernized. Digging deeper into the standard definition, it is clear that engineers have a built-in bias that favors consumption of natural resources. Unreconstituted "raw" nature, whether a forest or a seam of iron ore, is "obsolete." This bias stood in opposition to "modern." Land waiting to be used was obsolete until it had been "improved" and thus given value by human endeavor. (The English philosopher John Locke had long justified this belief for Americans.) Undomesticated land did not even deserve the designation "property." As wilderness, it was obsolete, with no intrinsic worth. Its potential was merely dormant. Anything so intrinsically useless deserved to be exploited for human well-being, profit, and the fulfillment of Manifest Destiny.

The power of this mental construct in shaping American history cannot be overemphasized. Cultural historian Daniel J. Boorstin writes: "The map of America was full of blank spaces that had to be filled. Where solid facts were scarce, places were filled by myths."[8] One symbolic geography, savage wilderness, was traded for another: the engineer's dreamworld of a fully mastered nature. In this view First Nature, whose natural forces restricted, enclosed, and trapped Americans, deserved to be conquered and exploited. The engineer's "Second Nature" would speed America's move toward paradise.

Engineers have always been dissatisfied with existing conditions. Hence Americans came to adore them. They promised a superior world around the corner, given the proper tools, money, and initiative. Historian Lynn White Jr. put into words what most Americans believed: "Engineers are the chief revolutionaries of our time. Their implicit ideology is a compound of compassion for those suffering from physical want, combined with a Promethean rebellion against all bonds, even bonds to this planet. Engineers are arch-enemies of all who resist the surge of the mass of mankind toward a new order of plenty, of mobility, and of personal freedom." White, who was not an engineer, concluded that, with engineers at the helm, America's built landscape was a "prime spiritual achievement."[9]

Engineers suddenly seemed to swarm over the American landscape. Geotechnical engineers asked mundane but essential questions about rock and soil mechanics: Is the terrain stable? Can it bear the weight of the project? Structural engineers focused on foundations and the steel, concrete, and masonry needs of new buildings. When twentieth-century civil engineers sent an entire superhighway through an urban neighborhood, the project involved the building of the roadbed, revetments, ramps, and bridges; the placement of water, sewer, and utility lines; and the removal of entire neighborhoods of homes and stores. The result is a public work. Private civil engineering similarly includes the creation of pipelines, shopping malls, housing tracts, and other large construction for industrial, commercial, and residential use. In the early 1970s environmental engineers (earlier called sanitary engineers) entered the picture to solve problems of sewage treatment, air and water pollution, and hazardous waste. Yet not until 1900 had there even been enough engineers to form a professional class between the traditional "mechanick" and the entrepreneur. The ranks of engineers, scarce in 1800, numbered in the hundreds in 1870; leaped above 230,000 by 1930; and soared upward of 800,000 by 1990.

Americans truly believed they lived in a new Age of Progress created in particular by engineers, who dirtied their hands with the nation's raw materials. The nation took on a new physical appearance that featured not only bridges that spanned America's rivers but also railroad viaducts like Starrucca in Pennsylvania, tunnels through Hoosac Mountain and under the Chicago River, and especially the transcontinental railroad. Nineteenth-century Americans put their greatest energies into laying railroad tracks—many miles a day—because doing so would fulfill their destiny. Twentieth-century Americans would thrill at engineering marvels like the Empire State Building and Boulder Dam, both built at record speed. This physical expansion of America depended upon extravagant visions of Second Nature. Engineers certified that a grand idea that could become a paper plan could lead to the construction of a lasting structure like New York City's Chrysler Building or San Francisco's Golden Gate Bridge. These were grounded virtual realities.[10]

Historian David M. Potter describes the prevailing optimism in his 1954 *People of Plenty: Economic Abundance and the American Character.* He sums up the new confidence in America's future as a continuous wealth machine drawing its energy from an abundant nature perpetually refreshed by invention:

The American Indians possessed, but benefited little from, the fertile soil which formed an unprecedented source of wealth for the colonists; that the colonists gained little more than the grind of their grain from the water power which made magnates of the early industrialists; that the early industrialists set little store by the deposits of petroleum and ore which served as a basis for the fortunes of the post–Civil War period; that the industrial captains of the late nineteenth century had no conception of values that lay latent in water power as a force for generating electricity, which would be developed by the enterprisers of the twentieth century; and that these early twentieth-century enterprisers were as little able to capitalize the value of uranium as the Indians had been five centuries earlier. The social value of natural resources depends entirely upon the aptitude of society for using them.[11]

Potter concluded that America's geography, made abundant by the engineer, is the most tangible feature of its history.

By the engineer's magical hand, Americans came to inhabit *infrastructure*—the built environment—instead of wilderness. Infrastructure became a habitable Second Nature, replacing the uninhabitable First Nature. This shift can be called America's big bang, expanding exponentially to envelop the nation's geography. Americans insisted that they had the right to seize the initiative from the natural world. An entire geography could be conquered and bypassed by a totality of physical grids, industrial systems, and urban infrastructures.[12] Webs of roads, canals, and railroads, joined later by intricate water systems, interstate highways, and metropolitan sprawl, now overlie nature. This initial infrastructure soon included electric power grids and natural gas lines, telephone and other communications connections, and all the other structures and connections that we inhabit. Dams, reservoirs, aqueducts, tunnels, and canals reshaped and dominated large geographical regions such as California's Central Valley—turning a desert into a garden. Engineers sought to make all things new—in journalist Joel Garreau's words, "to redefine human nature and society, eliminating every vestige of the stale, unoriginal and sentimental." The landscape was to be made up of roads and bridges, canals and railroads, skyscrapers, the automobile, and the shopping mall.[13]

Geographical distance and physical obstacles disappeared from view because of Second Nature. Cleveland and Pittsburgh competed not with their rural outliers but with distant Atlanta, Kansas City, and Denver. Dallas and Houston would have remained small hamlets without the automobile, the trucking industry, air travel, and air conditioning. America's major cities—New York, Boston, Philadelphia, Detroit, Chicago, and Los Angeles—no longer are mediated through state governments, since state boundaries often mean little in terms of commerce and populations. Indeed, in the second half of the twentieth century there was often more exchange between cities and the federal government than between cities and state governments. This national urban network emerged as the dominant material feature of the nation, this time a built environment created by a priesthood of engineers.

Engineering's move into the centers of power took hold during the Progressive Era, at the turn of the twentieth century. Engineers, idealized as archetypal "experts," would lead the nation into a planned industrial utopia. President Theodore Roosevelt and colleagues like forester Gifford Pinchot preached that the nation's natural resources, notably its vast forests and waterways, must be efficiently controlled, used, and conserved through engineering know-how. Herbert Hoover, who had first gained the public eye through his engineering feats, rode the profession's coattails into the presidency. For a time in the first half of the twentieth century it was widely held that the engineer was the most moral person in the nation: technological solutions could address social needs and ills.

Theodore Roosevelt's Progressivism worked out the general principles that Franklin Roosevelt's New Deal put into practice more than two decades later.[14] Indeed, one of the most powerful examples of a transformed national commons remains the Tennessee Valley Authority (TVA). Roosevelt's New Deal became the center of a vigorous ideological debate over the future of the nation's physical territory. The New Deal, as it reshaped America's infrastructure, has been called "the third American Revolution."

24A and 24B. Highly competitive Americans loved nothing better than a race. Speed meant the conquest of the nation's geography. Steamboats fulfilled the long-standing human dream of the reliable movement of people and goods upstream, an unheard-of phenomenon in 1800. And railroads increased speed dramatically, taking goods and people across challenging landscapes. Both of these Currier & Ives images also depict the technological conquest of night, a time when previously—recently—all human endeavors had slowed or halted. Not the least, both images are statements of technological endeavor that verified national self-consciousness.

A Revolution in Geographical Perception:
Engineering the Compression of America's Space

As late as the 1840s the United States still stood as a nation of "island communities" loosely strung across the wide landscape. Colonel John J. Abert of the Topographical Engineers wrote about the Plains and the West in 1849: "Unless some easy, cheap, and rapid means of communicating with these distant provinces be accomplished, there is danger, great danger, that they will not constitute parts of our Union."[15]

One technology was already set in place. It is difficult today to imagine the dramatic change Americans felt when the steamboat first allowed them to move *upstream* against the powerful currents of the Ohio, Mississippi, and Missouri rivers. Robert Fulton's *Clermont* ran upstream from New York to Albany in 1807, a remarkable 150 miles in 32 hours. The first steamboat launched in Pittsburgh, in 1809, made her maiden journey to Louisville in 70 hours. In comparison, in 1782 it had taken the French traveler Crèvecoeur 212 hours to make the same journey. Steamboats effectively pulled the nation into a network that stretched from the Atlantic Ocean to the largely waterless Plains.

Now a new technological system—the railroads—provided the solution to Abert's problem, and more. The virtues of the railroad soon became apparent. The rails, mind-bogglingly, went directly "as the crow flies." The railroad could climb tall mountains, leap yawning chasms, and ford raging rivers. The speed of trains was triple that of the finest horses and wagons, and the aptly named "iron horse" never tired. It was dependable in all seasons and all weather. The steam locomotive could plow its way through winter snowdrifts and blizzards, and sturdy rails and well-engineered beds kept trains from becoming trapped in muddy quagmires in the spring. One construction boss boasted, "Where a mule can go, I can make a locomotive go."[16] The natural world had become a bystander, subordinate to the technological juggernaut. Everyone marveled at the engineering miracles: cuts and grades, tunnels blasted with dynamite, tracks laid through seemingly impenetrable terrain, river crossings.

The new network of tracks made nature disappear as never before. Railroad systems dominated the maps of the day: the blank spaces between the islands came to be regarded as temporary geographical impediments. The head of the Union Pacific survey, General Grenville M. Dodge, remembered, "When you look back to the beginning at the Missouri River, with no railway communication from the East, and 500 miles of the country in advance without timber, fuel, or any material whatever from which to build or maintain a road, except the sand for the bare roadbed itself, with everything to be transported, and that by teams or at best by steamboats, for hundreds and thousands of miles; everything to be created, with labor scarce and high, you ask, under such circumstances could we have done more or better."[17] Americans, still amazed by the steamboat, were incredulous in 1820 when Robert Mills, a visionary construction engineer, claimed that they would someday ride a train from coast to coast in sixty days; by 1890 passengers were crossing in seven days. Railroad lines crisscrossed the East by the 1850s and began to bridge the West in the 1870s, binding together an otherwise impossibly expansive geography.

The push westward had the inevitability of Manifest Destiny. The transcontinental link—"gigantic thrusts into and across the heretofore unsettled domain"—may indeed represent the greatest technological conquest ever over the earth's geographical space, particularly considering what had existed only a few years earlier. When in 1869 the two track-laying parties of the transcontinental railroad met at Promontory Point north of the Great Salt Lake, the whole nation stopped to celebrate. A special 1876 centennial train run of 3,317 miles between New York City and San Francisco astonished celebrants with a record time

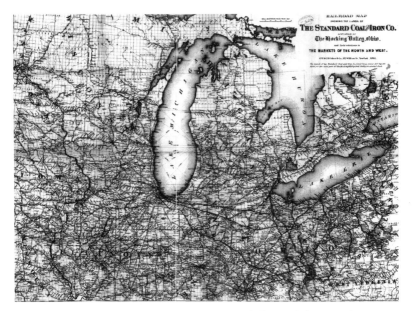

25. Railroad networks have been compared to the web that might be woven by a neurotic spider. This map of midwestern railroads demonstrated to Americans of the 1880s that their large geography had been conquered, even diminished, with virtually every small hamlet now connected to major metropolises like Chicago, Saint Louis, Cleveland, Toledo, Louisville, and Pittsburgh. Between 1880 and 1950 the railroad network probably surpassed the modern network of major highways and certainly surpassed currently available airline service.

of eighty-three hours and twenty-seven minutes, at an average speed of forty miles per hour. By 1890 72,473 miles of track lay west of the Mississippi. Even the desolate Great Plains were crisscrossed by a network of rail lines that looked like the work of a neurotic spider. By 1900 railroad tracks had reached 193,000 miles, peaking in 1916 at 254,000. The picture of America as Nature's Nation disappeared, in effect, under a web of steel tracks.

The American ambition was not merely to defeat nature's purposes but to redirect and reshape nature itself. It was just a matter of time until Americans succeeded in collapsing the nation's vast spaces and physical barriers into manageable pieces.

26A and 26B. These two images of urban America depict two extreme visions of what industrial America offered its inhabitants. On one side is the seemingly pleasant aspect of downtown Cleveland. On the other side are two Pittsburgh children pumping water that is suspect because privies were often adjacent to wells. Urban infrastructures could enhance quality of life and standards of living like nothing else, but reformers perceived a negative infrastructure that suggested squalor, degradation, and impoverishment. It is striking that both images feature children.

Through the networks of track, locomotives, and bridges, and reliable timetables that demanded the invention of national time zones, the railroad reigned over transportation from 1830 to 1950. The railroad, indeed, acted as the first comprehensive circulatory system of the middle-class world. Everything seemed to be connected to everything else.

The railroad revolution lasted for over a century before Americans shifted to personal automobile travel and freight moved into trucking. Often the new highway routes, like the Pennsylvania Turnpike, overlay existing railroad lines, grades, and tunnels. The interstate highway system and the airline network would certainly transform America's geography again, but they were only reinforcing patterns the railroads had already established.[18] The spacious nation was no more.

Dystopia

Had Americans made a bad bargain—a Faustian bargain—between the rise of Second Nature and the ensuing corruption of First Nature? In older established cities like Boston and

Philadelphia shops, parks, and individual homes had once of-fered pleasant livable neighborhoods. Even so, the natural or-ganic wastes of a crowded city, such as the tanneries' putrid smells and leftovers, were hard to bear. Beginning in the 1850s, however, a troublesome new list of wastes appeared: the sulfur, cyanide, and ammonia gases of industrial operations; kerosene smoke from lanterns; acid fumes from metal plating shops; every-where cinders and coal dust and the heavy particulates of black industrial smoke. Smoke had long been equated with progress: Pittsburghers took pride in listing fourteen thousand smoke-stacks in their region. City dwellers were not supposed to be put in "harm's way," but they were, and the effects of this gravely compromised Engineered America. Side effects of the factory sys-tem, the harm done to the commons and public health, should have induced negative cost accounting. In Pittsburgh, for ex-ample, an unnecessary typhoid epidemic in 1906–7 cost the city $540,000 in expenses and lost wages. And overall, typhoid cost Pittsburgh nine million dollars between 1883 and 1908, when a water filtration plant was finally completed.

A vast negative infrastructure had taken over the factory cit-ies. America's crowded worker housing, for example, ran into ganglia of pain, disease, death, and despair. Industrial slums intensified physiological problems: children deprived of sun-light developed rickets, their bones malformed by a bad diet, sick with smallpox, dysentery, and cholera because of dirt, ex-crement, and overcrowding. The backyard privy might adjoin a neighbor's private well, and a newly arrived immigrant family might be living in a basement amid foul seepage from the walls and muck on the floor. Bad drinking water and human waste seeped through deficient sewers or open drains. Garbage dis-posal was nonexistent. So was health care.

The built landscape may have generated a great deal of wealth, but it was not creating the livable society that had been promised. Sam Bass Warner reported that Americans were squandering the opportunity provided by the "great once-on-ly experiment" of technological innovation; they "endlessly failed" to build and maintain humane cities. Their failure arose, Warner said, from inherent flaws built into American industrial capitalism—flaws that promoted private profit-centered competition and innovative change without regard for community: "a safe, healthy, decent environment for everyone, regardless of personal wealth or poverty, success or failure."[19]

Nevertheless, infrastructure systems effectively removed the historic constraints of distance and physical obstacles. The natural terrain of a city receded into the background. On a regional and even a national basis, an entire geography was being conquered, bypassed by this totality of physical grids, industrial systems, and urban infrastructures. In its short history the United States had engineered a remarkable, widely admired, and widely imitated infrastructure, a technological layer upon the environment intended to improve the standard of living. By the mid-twentieth century highways and airlines raced across the rural countryside. Second Nature had taken hold—our steel-beamed, concrete-walled, and polystyrene world. Despite its defects, Second Nature appeared far more habitable than the original First Nature.

Consumer America

Maverick environmentalist and desert rat Ed Abbey hated consumerism. He saw the struggle against excess as a dead-serious battle for the American heart. And it was probably a losing battle. As Abbey trudged off to a canyon hideaway, he raged: "There comes a day when a man must hide. Must slip away from the human world and its clutchy, insane, insatiable demands."[20] To Abbey consumerism was not good enough to define American life. It was a single-minded juggernaut that constrained what remained of the nation's imagination.

Abbey had found a ready target. It is all too easy to critique consumerism and the public's gullibility as signs of a stunted dream. Americans have long measured a successful life by wealth—"the velocity of money"—and the continuous accumulation of things. In the 1840s critics like Ralph Waldo Emerson already warned that "things are in the saddle, and ride mankind."[21] And Abbey personified the contradictions that inhabited most American dreams: along with desert individualism he loved his clapped-out Ford pickup, cold beer in throwaway aluminum cans, and a mass market for his books.

Fulfillment of the consumer fantasy involves, in social critic Mary Gordon's words, "a transaction between the symbolic and the actual which many people believe is the most real thing there is, but which is, in fact, a sheer act of imagination."[22] Consumerism, in all its virtual guises, jumps to the forefront of our daily lives. Historian Harvey Green describes how material goods might satisfy the dreamworld of consumerism: "Artifacts are mute and could have multiple meanings for their users. . . . The material base of society—houses, artifacts, the visual world—is a key signifier of a culture's ideas, identity, and intentions."[23] To be surrounded by the clutter of the latest goods is thus liberating: the old family van is traded for an suv, which is only leased until the new, improved model comes out. Fuel costs bring forth a new craving, this time for the virtue of hybrid vehicles. The stereo from college days is transmogrified into the home theater. The 2002 transition from vhs to dvd was a triumph of consumerism, as was the switch from analog to digital television, and from a wired to a wireless world. Extravagant, conspicuous consumption, such as the $60,000 wine binge indulged in by some banking executives, depends on the ability to let one's imagination go rampant, but down the single narrow alley of desire. Gordon continues: "We refer to such activities as 'mental health purchases.'" But she admits that these are "cloud-cuckooland times." Like most virtual worlds the consumer dreamworld is distinguished by a quasi-religious fervor. Victor Lebow adds: "We convert the buying and use of goods into rituals. . . . We seek our spiritual satisfaction, our ego satisfaction, in consumption."[24] The mantra of my students in the 1980s and 1990s was, "He who has the most toys, wins." Having more is *being more.*

Is American History Best Told as the History of the Consumer?
The Triumphal Victory over Want

Consumerism has noble origins. Unspeakable scarcity, not abundance, defined most of past history. Well into the nineteenth century the typical European was a peasant tied down by few material goods—homemade plow, hoe, and spade; handhewn table, bench, and feeding bowl; perhaps an iron pot and plow edge from the local blacksmith. Peasant and family were devoid of hope for improvement.

Not until the eighteenth century did the combination of the Scientific Revolution and the Industrial Revolution slowly bring about a primitive consumerism. French philosopher Jean Jacques Rousseau called consumerism the willingness to "bear with docility the yoke of public happiness."[25] Precepts promoted during the Age of Reason marked the beginnings of the modern belief that human happiness could be measured by improved material life—better health, durable shelter, and satisfying work, rather than heavenly spirituality or military glory. Pain, suffering, poverty, and deprivation were not humanity's eternal lot. Science uncovered First Nature's secrets. Industry transferred its riches into the mass production of material goods. For the first time in human history the artifacts necessary to live the good life became available at low cost to a general public.

Early on Americans defined progress in part by the variety, palatability, and convenience of their diets. A family's consumption of beef became a symbol of its prosperity. Beef made up two-thirds of the typical meat diet, pork a quarter, mutton or fowl the rest, with virtually no fish or shellfish. At an Ohio inn in 1807 a traveler could sit down to "good coffee, roast fowls, chicken pie, potatoes, bread and butter, and cucumbers."[26] Other roadside menus featured enormous portions of pork, beef, and game, as well as turnips, peas, beans, apples, cherries, strawberries, and melons.

Rural life was at its best in America, and not only in terms of food. By any measure of the day Americans had developed the best standard of living in the world in the six decades between 1800 and 1860. The everyday spaces of home and neighborhood, regular travel to town, the feel of the familiar axe or plow or kitchen knife or broom, even furnishings—all established a consistent sense of well-being. Poverty meant the lack of these amenities and an uncomfortable closeness to the rough environment of First Nature. A family was judged lower-class and unsuccessful if it had little or no furniture, no artificial light (even candles were costly), no indoor toilets or running water, and if all its members, male and female, worked in the fields and barn. Americans enthusiastically promoted the "ideology of domesticity" as a sign of upward mobility. Many young farm-bred women, by the mid-nineteenth century, began to see themselves filling the desirable position of "housewife"—full-time domestic worker—rather than field worker. Success meant better buildings, tools, and artifacts that kept the environment at a distance. A reliable surplus of corn or wheat or livestock allowed the frontier farmer to acquire items that belonged only to the well-off in Europe: tea, pepper, chocolate, cotton goods, gunpowder, tobacco, ironware, boots, hats, a stove, window glass.

Alexis de Tocqueville, in his 1835 classic *Democracy in America*, observed America's fixation on material well-being: "It is odd to watch with what feverish ardor the Americans pursue prosperity and how they are ever tormented by the shadowy suspicion that they may not have chosen the shortest route to get it."[27] He warned that American materialism was "a dangerous disease of the human mind."[28] In 1851 the novelist Herman Melville had Captain Ahab say, "All my means are sane, my motive and my object mad."[29] Thoreau's 1854 *Walden* remains a classic diatribe against commercialism and materialism. The ambiguity that some Americans felt was reflected in the economist Francis Wayland's popular 1838 book, *Elements of Political Economy*. Wayland complained of reckless, self-indulgent consumerism but also praised the rising tide of labor-saving devices and cheaper goods. The working poor *could* enter the middle class. The Enlightenment's secular optimism had seemingly been fulfilled in America, a vision of worldly salvation, a heaven on earth.

Historian Henry Bamford Parkes describes the novelist Mark Twain's late-century struggle to understand the new consumer world that had been flung upon him:

> Twain can be cited as a case study of how the agrarian American submitted to capitalism. He had all the characteristic virtues of his agrarian background; a natural democrat, with the fundamental American respect for the rights of all human beings, he despised pretense and sham, and hated injustice and exploitation. At the same time, he had no coherent social philosophy; his political affirmations were instinctual rather than reasoned. He had no capacity for abstract thought and little respect for intellectual speculation, and his opinions, in spite of his homely and realistic common sense, were often remarkably naive. Moreover, he was personally as eager as most other Americans to achieve material success and to discover some easy way of making money. Transplanted from the frontier to the East, he could neither accept nor repudiate this new environment.[30]

Parkes adds, "He [Twain] saw the dishonesty and the exploitation that accompanied the rise of capitalism; but he had no

alternative social doctrine to propound, and he was too honest merely to condemn the robber barons without recognizing
that they were doing what other Americans would have liked to
do if they had the opportunity." Mark Twain, who in his later
years would succumb to financial speculation in an immediately
outdated new printing technology, fell into deep melancholia.
Parkes concludes that he "was, in a sense, broken by capital-
ist America."[31] Frederick Jackson Turner, historian of the frontier, feared by 1900 that the new Consumer America retained
no connections with the dramatic frontier expansion that had
shaped the nation's original powerful ideals and values. Factory
America was a different country, he said, from frontier America.
As early as the 1890s most Americans realized that consumerism had created a distinctive society that now had a life of its
own, seemingly beyond their ability to control.

American Entrepreneurs Take Charge

By the 1890s Americans had added to their roster of heroes
the inventor-entrepreneur, who joined the engineer, the fron-
tiersman, the farmer, and the military leader. Andrew Carnegie,
the most literate of his peers, became a spokesperson for "The
Gospel of Wealth." Material success provided the means for human advancement because it raised the standard of living not
for just a few but for all. Carnegie described three basic principles that, if adhered to, might allow American prosperity to
advance forever: the divine right of private property, the laws
of competition, and the stewardship of wealth. The connection
between Protestantism and the spirit of capitalism made productivity a religious calling. The enemy was the untamed environment of First Nature. Carnegie's contemporary, psychologist and
philosopher William James, concluded that only by struggling
against the forces of nature would the human spirit have a fighting chance of safety against an indifferent physical world.[32]

One of the most perceptive analysts of the new consumerism, Thorstein Veblen of the University of Chicago, worried about its uncontrollable momentum. In his 1899 classic *Theory of the Leisure Class* he wrote that the luxuries of an earlier day become the next generation's entitlements; these hoped-for entitlements then become the necessities of yet another generation. Anticipating modern environmentalism, Veblen strove for a holistic image that would avoid falsely separating humanity from larger environmental forces. Veblen acknowledged that Americans enjoyed industrial efficiency, productivity, and technological improvements, but their enjoyment came at the price of contention, distrust, and chicanery. Human nature had taken a turn for the worse. In a vivid precursor of ecosystems analysis Veblen concluded that because humanity, technology, and the environment inescapably formed an organic whole, a change at any given point initiated changes everywhere else.[33]

Nevertheless, the voices of Veblen and other critics did little to distract most Americans from their joyous embrace of consumerism. Social Darwinism suggested an updated environmental determinism in which individuals had little choice but to conform to their newly industrialized surroundings, now based on technology (mechanistic Second Nature) rather than nature (organic First Nature). Remember the slogan of the 1933 Chicago world's fair: "Science Finds—Industry Applies—Man Conforms."

Consumerism took off after World War II. Robert Heilbroner, twentieth-century economic philosopher, declared consumerism America's overriding adventure: "We watch the unfolding of material life as it constantly expands the limits of the possible."[34] An accelerated flow of goods would be cumulative and self-sustaining—a perpetual-motion growth machine. Social critic David T. Bazelon concluded: "The industrial system is a wonderful wealth-machine, and especially here in the United States, in its advanced state. But I want to be clear and very firm about what it is *not*; it is *not* a human society."[35] More than a century earlier Veblen had coined the phrase "conspicuous consumption," decrying this practice. Writer Robert Frank defined contemporary consumerism as "luxury fever": "across-the-board increases in our stocks of material goods [that] produce virtually no measurable gain in our psychological or physical well-being."[36] Juliet Schor identified "the overspent American": "Today [rather than simply keeping up with the Joneses] a person is more likely to be making comparisons with, or choose as a 'reference group,' people whose incomes are three, four, or five times his or her own. The result is that millions of us have become participants in a national culture of upscale spending."[37] She terms this the dumbing down of American idealism, to the extent that Americans define themselves either in terms of consumerism or nothing.

Sometime during the twenty-first century the costs of consumerism will shoot past its benefits. Advocates of so-called sustainability do not deny the human benefits of growth, but they do insist on the recognition of its total costs. They also raise serious questions about the uncontrolled growth that is the outcome of ill-formed marketplaces. In 1986 17 percent of the United States' GNP went unrecognized as the costs of air and water pollution, together with long-term degradation. Looked at another way, between 1980 and 1986, even if long-term degradation were taken out of the picture, environmental costs took 7 percent out of the GNP, which was otherwise claimed to rise by 11.6 percent. In 1989, after the economist Herman Daly and the ethicist John Cobb reviewed the previous twenty years, they concluded that the link between increased industrial production and increased human welfare had become progressively weaker.[38] A satisfying society is attentive to personal rights, family stability, health, and safety, as well as respect for nature.

A Tough Assignment: In Which Ecosystem Do You Live?

Human history, as Second Nature, is the outcome of the technological layering of farms, factories, cities, superhighways, suburbs, and shopping malls upon First Nature. Nevertheless, we inescapably coexist with the First Nature of forests, soils, water, climate, and other energy systems. Nature is still the precondition of our existence.

Do we belong to our immediate geographical place, or do we as individuals inhabit a lumpy, distorted, stretched world where many parts are missing?

Today we often have more connections with distant markets than with our local landscape. Our locus of personal space becomes more difficult when we realize that our breakfast banana comes from Guatemala, our clothing from China, our cut flowers from Columbia, our computer parts from Taiwan, and our automobile from Korea. Our personal artifacts seldom reflect our "home" place but are pulled to us from elsewhere. We act as magnets. Without the slightest inkling, we might have more material investment in another people's ecosystem than we do in our own. It becomes increasingly difficult to know which ecosystem is interacting with which culture. Can we cherish the places our bananas, clothing, and flowers come from? If we see a place only in terms of its industrial or consumer utility, or only its scenic or historic value, we're missing the point.

Sara Terry, a writer for the *Christian Science Monitor*, was assigned in the spring of 2003 to "eat locally" for a week—to buy and eat food produced in her own home place. Since she lived in Los Angeles, she thought, surely this would be a cinch: farmer's markets everywhere all year-round, specialized supermarket chains like Whole Foods, and a look on the Internet for other local sources. Since it was early April she found local broccoli, cauliflower, mushrooms, peas, carrots, onions, potatoes, and all kinds of greens. She located local fruits like oranges, grapes, strawberries, and dried fruits and nuts. Best of all, she met wonderful farmers and expert salespeople, an unexpected benefit of exploring local possibilities. Another writer for the same newspaper, Jennifer Wolcott, learned that the Japanese call this practice *teikei*, or "putting the farmers' face on food."[39] Such a connection can uncover a local rural landscape.

But then things got harder for Terry: no bread or pasta, since Southern California produces no wheat; no butter, chicken, or beef, and she was not a fish person. She was forced to expand her definition of "local" away from her old sense of place—someone living in Los Angeles typically places themselves in an area that stretches 120 miles south to San Diego and 200 miles north. But Terry's stretch for chicken, milk, butter, and cream, as well as any lunchmeat, like sliced turkey breast, meant Northern California. Beef meant Nebraska; lamb meant New Zealand. Terry found that her search took up a good deal of extra time and thought (not to mention money). But many of the results were positive, as her chef friend Evan put it: "I think eating locally starts as an abstract set of values which you want to conform to in your own life. . . . . But then you get completely seduced by the [superior] tastes of what you're consuming. . . . What you're consuming when you eat locally is so much more vivid in flavor and in meaning."[40]

Jennifer Wolcott, on the same assignment in her home territory around Boston, Massachusetts, found local grazing much more difficult. The winter menu might be limited to stored carrots and parsnips; other seasons would offer fresh local fruits and vegetables. But she noted that most people "would rather not wait until summer for green beans—or any other food for that matter—and they don't see why they should if these foods are always available and affordable."[41] Wolcott searched the Boston area in April, at the same time as Terry, but she found rhubarb from the Netherlands, mangoes from South Africa, and grape tomatoes from Chile. Other foods came from Costa Rica, the Dominican Republic, and of course California. She did find beets and parsnips from Vermont, Macintosh applies from upstate New York, and shitake mushrooms grown in Massachusetts. Fresh fish could be allowed if she extend-

27. No entry into the wilderness took on more significance than the Cumberland Gap, which opened the Kentucky region. The original title of the first version of this iconic painting was *The Emigration of Daniel Boone*, and it featured not anonymous settlers of an endless West but, on the white horse, Boone's wife, Rebecca, and behind her, their daughter, Jemima. At Boone's left is Flanders Callaway, Jemima's husband, wearing the coonskin cap that was Boone's signature. The second version was repainted to emphasize the risk to the adventurers. The rocky cliffs are narrower, the storm clouds darker. But Boone remains in a bright yellow homespun outfit, as if daring the elements and the enemies of America's rightful expansion into "empty land." A shaft of bright sunlight breaks through the dark clouds. The first work was done in Paris, the second in New York City. George Caleb Bingham did not create vast landscape canvases, instead depicting ordinary Americans as they serenely, and joyfully, entered the midwestern wilderness. If anything, the landscape, with its blasted trees, is picturesque, a middle ground between sublime and garden park.

ed her territory into the deeps of the Atlantic Ocean and did not think about the serious overfishing of the Outer Banks. The buyer for the local natural foods store, the pricey Bread & Circus, told her that her experiment, if strictly applied to early Massachusetts spring, would leave her "really bored and really hungry."[42]

Belonging to one's home place is a tremendous opportunity, but belonging to one's ecosystem is a tremendous challenge.[43]

Triumphal America: *God's Exceptional Nation*

For better or for worse, triumphalism is at the core of American history. Americans deliberately fabricated this self-conscious myth, which has always served as our dominant virtual reality. Call it Manifest Destiny. Call us the Redeemer Nation, Exceptional America, the Final Superpower. Regardless, Triumphal America is in effect an exaggeration of Manifest Destiny, which itself is an exaggeration of Puritan redemptionism. Triumphal America combines Engineered America and Consumer America under the umbrella of Manifest Destiny, a philosophy built on the back of America's benevolent geography.

The pace of American expansion for once surpassed the imagination. The success of Samuel F. B. Morse's magnetic telegraph invented a transcontinental unity. Railroads seemed to be a divine gift that would compress America's broad spaces. As America's geography thus contracted, according to historian Frederick Merk, American hubris expanded.[44] Theodore Roosevelt, asserting a jingoistic moral authority propounding that the United States deserved to create a worldwide empire, backed it up with force. Woodrow Wilson's Fourteen Points sought to make the world safe for American-style democracy. Wilson once told an audience that "America had the infinite privilege of fulfilling her destiny and saving the world."[45] And

28. This garish and awkward painting says it all, combining obvious myths of
frontier expansion, of America's national territory waiting for settlement. The
motifs meld individual success with Manifest Destiny. The Goddess of Liberty,
with the Star of Empire as a tiara, soars over the landscape, leading the way.
Her left breast demurely covered but mysteriously fastened, she carries a book
(either a law book, representing democracy, or a school book, representing enlightenment) and a reel of telegraph line (technology). In full retreat are buffalo, Indians, and even a raging bear (lower left), symbols of the untamed West.
Also featured, at center bottom, are hunters and miners, the rough vanguard of
civilization. At the lower right corner oxen pull the "sacred plow." The speeding
stagecoach nearly trips the goddess. The immigrant Conestoga wagon appears
to struggle westward under a heavy load. Three railroad lines, their trains ready
to overstep their rails, tell of the inevitable momentum of technology. On the
right horizon is a stable commercial city, hinting at New York City's Manhattan
Island. Everything is suffused with a rose-colored hue: light from the East will
brighten the West. On the western horizon mountains and storm threaten the
scene, with the Pacific coast barely in view, no doubt to be overwhelmed in due
time. John Gast's painting was quickly made into a chromolithograph by George
A. Crofutt and sold by the thousands, both reflecting and shaping the postwar
American mentality.

today's post-Soviet hegemony is touted as the fulfillment of U.S. righteousness and superiority, based on the combined victories of capitalism and democracy.

Manifest Destiny

In 1845 journalist John O'Sullivan coined the phrase "Manifest Destiny" in a Texas political publication.[46] A politician, adventurer, and scholar with potent personal charm, O'Sullivan promoted his vision of the nation's illimitable purpose with missionary zeal. O'Sullivan put into words a notion that had already been in the public mind for decades: the belief that Americans were destined "to overspread the continent allotted by Providence for the free development of our yearly multiplying millions."[47] In the words of historian Anders Stephanson, "Manifest destiny, history as revealed in the utopian space of America, would be *managed destiny*."[48] Americans became fiercely loyal to this determinism—to the inevitability of Manifest Destiny.

One dimension of this mentality was immediately clear: America's spectacular geography—its First Nature—was seen as empty space, in the biblical sense of the chaos that preceded the earth's creation (*tohu* and *bohu*). This divinely ordained opportunity freed Americans to transform this chaos into the rational order of private property, markets, and capitalism. The *Albany Argus* opined in 1845 that Americans who joined the Western migration became "peculiarly colossal" in their dedication to geographical destiny: "Does not this inspiration spring from their extraordinary country? Their mighty rivers, their vast sea-like lakes, their noble and boundless prairies, and their magnificent forests. . . . . To live in such a splendid country . . . expands a man's views of everything in this world. . . . .[America offers] great enterprises, perilous risks, and dazzling rewards."[49] In 1847 the *New York Sun* anticipated the dog-eat-dog philosophy of Social Darwinism: "He [the American] is carnivorous—he swallows up and will continue to swallow up whatever comes in contact with him, man or empire."[50] The *New York Post* editorialized, "Asia had her day, Europe has had hers; and it remains to be seen whether the diadem must not first be worn by the new world, before it reverts again to the old."[51] The continent was "a new earth for building a new heaven."[52]

After the annexation of Texas, O'Sullivan urged, "Yes, more, more, more . . . till our national destiny is fulfilled and . . . the whole boundless continent is ours."[53] War hawks demanded that all of Canada be integrated into the United States. Michigan senator Lewis Cass insisted that "Michigan alone would take Canada in ninety days."[54] Weak nations like Spain and Mexico, it was said, did not deserve sovereignty over Western territories; the continent belonged to Americans, instead of the wily and colonialist British and French. This view had already justified the Louisiana Purchase and the acquisition of Florida, Texas, Oregon, and California. Visionaries looked forward to an American Empire. And politicians were not to be outdone by journalists. Illinois congressman John Wentworth told his colleagues on January 27, 1845, that "the God of Heaven, when he crowned the American arms with success [in the American Revolution] . . . only designed them as the great center from which civilization, religion, and liberty should radiate and radiate until the whole continent shall bask in their blessing."[55] This was no lunatic fringe; it included President John Tyler and his secretary of state, John C. Calhoun.[56]

The tenets of a triumphalist ideology are hypnotizing: the power of science and technology is unlimited, nature's purpose is to be tamed to serve humanity, progress is inevitable, and enlightened humans like Americans represent the pinnacle of all creation. In the words of 1960s counterculture spokesman Stewart Brand, "We are as gods and might as well get good at it."[57] Triumphalism also depends upon environmental imperialism: we can twist nature any way we wish, turn the world into our artifacts. First Nature is swallowed up by Second Nature.

What are the specific features of this virtual reality, which sets the United States up as the world's exceptional nation?

Environmental superiority, because of our abundant food, natural resources, and land carved out of the wilderness;

Technological hubris, because of our vaunted innovation and productivity;

Individualism, based on the natural right to personal autonomy, largely defined by private property, as guaranteed in the Constitution and the Bill of Rights;

Corporate capitalism, dedicated to private wealth and consumerism;

Military hubris, based on the belief that we have (almost) "never lost a war" and the fact that our wars are usually fought on foreign soil (except the American Revolution, the War of 1812, the Civil War, the Indian Wars, and—150 years later—9/11);

Indifference and hostility to alternative ideologies and lifestyles (Woodrow Wilson said of World War I that it was fought "to make the world safe for democracy," American-style);

Passion for control and dominance, reflected in the belief in America as the "Redeemer Nation" and our powerful reformist tradition, jingoism, and national hubris;

Belief in the United States as the fulfillment of history, the embodiment of the final stage in human development, evolution, and progress.

*Blessed with Moral Virtue, Americans Obey a Different
Set of Laws: The Mentality of Exceptionalism*

The French visitor Alexis de Tocqueville was among the first to describe American exceptionalism.[58] The United States, it was believed, was not like the Old World but had been chosen by God to fulfill its special destiny, dominion over the entire earth. The old rules were obsolete: "Rapacity and spoliation cannot be the features of this magnificent enterprise . . . because we are above and beyond the influence of such views. . . . We take from no man, the reverse rather—we give to man."[59] America's millennial picture featured a continuing crusade of good against evil. There was no middle ground or Third Way: you were for or against America. War was infinitely preferable to a fallacious peace. The servant of God would be an occupying army that guaranteed a utopian future.[60] This apocalyptic flavor shows up in the famed 1862 anthem, "The Battle Hymn of the Republic":

> Mine eyes have seen the glory of the coming of the Lord:
> He is trampling out the vintage where the grapes of wrath are
>   stored;
> He hath loosed the fateful lightning of his terrible swift sword;
>   His truth is marching on.

> In the beauty of the lilies Christ was born across the sea,
> With a glory in his bosom that transfigures you and me:
> As he died to make men holy, let us die to make men free,
>   While God is marching on.[61]

Even such an open-minded liberal as Horace Bushnell concluded that human progress must include apocalyptic warfare against the forces of evil. True Christians, as God's instruments, were expected "to assert themselves to the utmost, in the use of all proper means, to suppress error and vice of every kind, and promote the cause of truth and righteousness in the world."[62] Religious historian Ernest Lee Tuveson describes Bushnell's jeremiad as a call not only to a crusade but to a "great jihad."[63] Theologian Samuel Harris wrote in 1870: "Any epoch in the progress of Christ's kingdom is liable to encounter violent and bloody opposition, and the advancement of Christ's kingdom may be in the midst of revolution and convulsion. In reference to this our Saviour said, 'I came not to send peace, but a sword.'"[64] Even the supernatural world—a heavenly virtual reality—could be appropriated for American interests. Herman Melville mused: "We Americans are the peculiar, chosen people. . . . God has predestined, mankind expects, great things from our race; and great things we feel in our souls. . . . We are the pioneers of the world; the advance-guard, sent on through the wilderness of untried things, to break a new path in the New World that is ours."[65]

Frederick Jackson Turner designed his compelling "frontier thesis" of American history in a speech made to historians meeting at the Columbian Exposition—the 1893 world's fair in Chicago. Turner spoke of waves of frontier settlement, moving continuously westward into "empty land," a phenomenon that gave Americans their personal sense of individualism, enterprise, and optimism, and notably their democratic institutions. Western expansionism made America exceptional, he said, granting the country a uniquely triumphal history, one that set them apart from all other cultures. The reach to the Pacific, he said, completed the creation of the United States. To others, however, Turner wasn't ambitious enough.

Rampant jingoist Josiah Strong pleaded, "My plea is not, Save America for America's sake, but Save America for the world's sake." Americans were a privileged Anglo-Saxon Protestant people who deserved dominion over the entire world. The spread of Americanism, said Strong, was foreordained: a "universal triumph is necessary to that perfection of the race to which it is destined; the entire realization of which will be the kingdom of heaven fully come on earth."[66] In 1900 Senator Albert J. Beveridge boomed before his fellow senators that God had prepared Americans as "master organizers of the world to establish system where chaos reigned." The Kingdom of God was not some unearthly realm that would come after the Apocalypse but a new human society built on the shoulders of America's new and unsullied geography. Beveridge declared that God "has marked the American people as His chosen nation to finally lead in the regeneration of the world. This is the divine mission of America." He added, "We are trustees of the world's progress, guardians of its righteous peace."[67]

By the opening of the twentieth century all the features of Triumphal America had been set in place. Theodore Roosevelt then reconstructed Manifest Destiny. Working closely with Secretary of State John Hay, Secretary of War Elihu Root, Senator Henry Cabot Lodge, and naval theorist Alfred Thayer Mahan, Roosevelt embraced an expansionist vocation that was not focused on the simple acquisition of a worldwide network of colonies, in the image of Britain and the European powers.[68] Rather, Roosevelt firmly believed that America's Manifest Destiny—once geographical expansion had reached the Pacific, and the frontier had closed—would naturally evolve overseas. Roosevelt's first implementation of this expanded vision was to secure the United States' status as a superpower through building a world-class fleet, as instructed by theorist Mahan. Before he was even president, he had pressured the reluctant William McKinley to show off the U.S. Navy's mastery of the seas during the Spanish-American War. Mahan's classic military treatise, *The Influence of Sea Power on History*, was not only devotional reading for Roosevelt but was later well-thumbed by German and British naval leaders in both world wars. As analyst James Chace writes: "Mahan was a frank imperialist, though he was not particularly interested in acquiring and administering territory. His aim was strategic."[69]

During the twentieth century the drive to save the world pushed U.S. entry into World War I in 1917 and stood behind Woodrow Wilson's later doomed support of the League of Nations. The "Good Neighbor" policy sugarcoated U.S. dominance, including through military force, in Central and South America. Theodore Roosevelt's cousin Franklin D. Roosevelt carried forward the former's internationalization of Manifest Destiny by calling upon the American people to propel "their righteous might" to mobilize the nation for World War II. And peacetime would bring the Marshall Plan, "a monument to American idealism."[70]

In peacetime 1948, according to Isabel Cary Lundberg, the world's population invented the virtual America it desperately sought to emulate: "the wrist watch, fountain pen, cigarettes, flashlight, chocolate bars, chewing gum, cameras, pocket knives, pills to kill pain, vaccines to save lives, hospital beds with clean sheets, hand soap and shaving soap, gadgets and gewgaws of every description, the jeep, the truck, and *white bread*. Very few Americans, picking and choosing among the piles of white bread in a supermarket, have ever appreciated the social standing of white bread elsewhere in the world. To be able to afford white bread is a dream that awaits fulfillment for billions of the world's population. To afford it signifies that one enjoys all the comforts of life."[71] Franklin Roosevelt is said to have claimed that he could overwhelm the Soviets if he could place a Sears, Roebuck catalog in the hands of every Russian. Consumerism merged into triumphalism.

After the collapse of the Soviet Union in 1990 the virtual reality of American triumphalism could be put fully into practice. This was such an unexpected surprise that Americans are still learning how to cope with their success. By 2002 the Russians had publicly acknowledged that the United States was the world's sole superpower in a new world order. Dimitri Trenin, a Moscow foreign policy expert, concluded, "We are managing the light of a [Russian] star which has been dead for ten years."[72] American triumphalism took a moral high road that promised worldwide individual liberty, democracy, and the personal right to a better standard of living and quality of life. The matrix of consumerism, the marketplace, and capitalism had helped induce—it was fervently believed—the climax of human history.

Two million people jammed themselves into New York City's Times Square on the evening of December 31, 1999, to join the countdown toward the ball dropping to mark the new year 2000, the opening of the third millennium. That venerable newspaper of record the *New York Times* editorialized, "Times Square will become, in effect, the virtual town square for a nation and indeed for a world reshaped by revolutions in communications, transportation, medicine, commerce and perhaps most forcefully by America's century-long act of self-invention."[73] One of the *Times*'s competitors in shaping the national agenda, *USA Today*, observed: "If we have conquered anything, it is time: weekend trips, 24-hour news channels, one-hour photo labs, 30-minute pizza and instant-cash ATM machines. We have obliterated time with microwave ovens, VCRs, cell phones, fax machines, books on tape—all designed to unshackle us from the burdensome, mundane tasks that consume our precious lives, and to allow us to experience more leisure and free personal time. How about it? Do you have more free time?"[74]

Summing up, the *Times* editorial told a story of historic triumphalism that had been jump-started by the founding fathers in 1776, "its founding idea to guide the world away from totalitarian rule and economic ruin." The *Times* continued: "The idealistic desire to make the world over is the deepest mystery of the American character and our signature national trait." This had been the dominant myth of the twentieth century: "It was a century in which freedom triumphed and generosity became a global ideal because of the most distinctive of all human inventions, a society based on the values of political freedom, economic opportunity, individual worth and equal justice." This exceptionalism was in the process of being fulfilled: "It is that invention that must be preserved and celebrated above all others, for it is our national treasure and the world's." The *Times* added that the United States had achieved global hegemony by the end of the century not so much because of its nuclear arsenal or containment of the Soviet Union but because of "the American economy and the worldwide influence of the popular culture made possible by that economy."[75] Commentator and professor Francis Fukuyama was ebullient: "That's it! We were in the final, and we just won! The United States is the last superpower! Now the world will want to adopt not just our political philosophy, democracy, but our economic system, free-market corporate capitalism, as well."[76] Fukuyama created an uproar when he proclaimed "the end of history," brought about by the superior virtue of capitalism over all other ideologies. However, British critic Godfrey Hodgson discovered in Fukuyama "an essentially Candidean standpoint. All is for the best in the best of all possible societies."[77]

Timothy Garton Ash, a member of think tanks based at Oxford and Stanford, asks, "What do you think of America? Tell me your America and I'll tell you who you are." Among Europeans, for example, Ash finds an America seen as "a dangerous, selfish giant, blundering around the world doing ill, and as an anthology of all that is wrong with capitalism." America, he says, "is part of everyone's imaginative life, through movies, music, television and the Web, whether you grow up in Bilbao, Beijing, or Bombay. Everyone has a New York in their heads." Ash concludes that America's worldwide influence is based in the virtual reality it has created about itself, its "soft power," rather than its military or economic power.[78] Australian policy analyst Owen Harries acknowledges, "We know the American landscape about as well as we know our own: the prairies, the Manhattan skyline, the white spires and fall colours of New England, those dangerous small towns of the Deep South—to some degree they are all part of our inner landscape." He adds: "We have seen perhaps a thousand American movies. . . . We have seen hundreds of American sit-coms. We know the words of dozens of American popular songs. We have read

Hemmingway [*sic*] and Fitzgerald and Steinbeck and Bellow and Updike. . . . By now we are even familiar with American idioms and regional accents."[79]

The Triumphalist Myth

Think of global geopolitics as cyberspace—an ether of pure chaos, anarchy, no rules, and inherent uncertainty. Think of nation-states as virtual realities floating in this cyberspace. Or as Peter Gowan puts it, "States, the principle agents of the international system, can be treated as so many black boxes or billiard balls." American hegemony, Gowan writes, requires "an economic, social and cultural leadership, resting not just on military force, but on an ideological ability to impose on allies and even adversaries the images and idealizations of the hegemonic state as universal values. Who could doubt the grip of the 'American Dream'?" Now think of the same cyberspace, but this time filled with multinational corporations instead of nation-states—Exxon-Mobil, Bechtel, Halliburton, Microsoft, Holland's Shell Oil, Germany's Siemens, and Japan's Sony—whose identities depend upon visions of growth and profit. Gowan looks at America's economic security as "the commanding vision of the architects of the American century, from Elihu Root through Stimson and Acheson to the Rockefellers, who believed America's surplus capital could transform and knit the world together."[80] Gowan wonders whether the nation-state still holds sway or has been replaced by multinational corporate black boxes.

Take the same cyberspace, and now fill it with individual-istic Americans laboring as consumers. The commentator John Lukacs argues that the modern age is the age of money: "More than ever, money is an abstraction, due, in part, to the increased reliance on entirely electronic transactions and on their records. Income is more important than capital, quick profits more than the accumulation of assets, and potentiality more than actuality." Real cash no longer changes hands; potential assets minimize real assets. This virtual world has taken on greater significance than material wealth—"the increasing intrusion of mind into matter"—despite the fact, says Lukacs, that this is shifting "at a time when philosophies of materialism are still predominant . . . [which] reflects the mental confusion of our times."[81] The outcome of wealth has become total personal privacy, total insulation, regardless of how lonely and boring it might be. Think Howard Hughes and Elvis Presley.

Think also of the Internet in cyberspace. Here too American triumphalism is rampant. The Internet is not as stateless as some advocates had hoped. Indeed, in many respects it is the new version of U.S. territory, and not simply because it was originally used to enable transfers among American universities involved in the defense business. As journalist Steve Lohr noted in early 2000, the Internet is an American colony—the first global colony.[82] Its dominant language is American English. Internet specialist Don Heath adds, "If the United States government had tried to come up with a scheme to spread its brand of capitalism and its emphasis on political liberalism around the world, it couldn't have invented a better model than the Internet."[83] On the other hand, Siemens executive Gerhard Schulmeyer insists somewhat hopefully that "reasonable people understand the Internet is a technology platform, not some form of American imperialism." However, he admits its pressures: "The Internet moves rapidly and is a disruptive technology that undermines institutions of all kinds, companies, and trade unions."[84] The Internet is difficult to control, as the Chinese have discovered. It tends to be individualistic and decentralized, volatile and destabilizing.

Engineered America, Consumer America, and Triumphal America are three working pictures of the American experience. They shaped the past and function even more powerfully today because they are increasingly interlocked. If we can envision these as overlapping virtual realities, we see that they are more than myths or symbols or metaphors: they are simulations that became realities. They inhabit their own real worlds, enjoying enormous power as seemingly more real than material existence. We run these simulations repeatedly in our heads, our writings, our actions. These three visions of America are species of the imagination that seemed beyond real-world fulfillment yet led to one of the most surprising outcomes in world history: global U.S. cultural, economic, and political preeminence.

5

Finding Authenticity

Inhabiting Place in America

An earnest young sociologist worked her way into an Appalachian hollow by tracking down people in little squares marked on a map—"dwellings on intermittent streams": "I finally made it to one and knocked on the door. A young mother appeared. 'You must live here,' I said, pointing to the map. She looked at me with a quizzical expression. 'Reckon I hain't. I's livin hyur,' and she gently stamped her foot on the floor."[1] One is tempted to smile, but at whom, the mother or the sociologist? One of any person's most valuable assets is a highly individualized sense of place. We discover a special bond with its concreteness, exuberance, and quality. The heart of an authentic America is less in the big picture or larger philosophies than in the specific sites of vivid human experience.[2]

The young Appalachian mother saw the sociologist's map as only someone else's piece of paper, totally irrelevant to her place. Her place was "here," in her centered existence, not "there," on one of the interchangeable little squares on the abstraction we call a map. She stamped her foot on her concretely palpable reality that secured her own personal identity. Her assertion was not ignorant; her spontaneous intuition told her that the well-worn wooden plank floor was an absolute source of her identity. Her place was honest, unspun, undiluted, longstanding, rooted, and human. Her attachment was concrete and passionate. She vigorously refused to surrender to the printed information on the map. Indeed, she made it clear that to her the map was meaningless. It belonged to some strange other world. The sociologist left, confused by the bad fit between the science of her map and the dramatic claim of the mother.

Place is unique to each of us, a means by which, in geographer Donald Meinig's words, "we pierce the infinite blur of the world and fix a piece of our environment as something distinct and memorable."[3] Pascal wrote of his fright at how irrelevant humans are to the totality of the universe: "The eternal silence of these infinite spaces terrifies me."[4] Human societies for millennia have rushed to overcome primeval confusion through creation stories, initiation rites, and scientific observations. The French philosopher Michel Foucault concluded that we couldn't live inside the void of unstructured space. Instead we must live inside concretely tactile places that are "absolutely not superimposable on one another."[5] They are unique and irreducible. As individual selves we cannot be a "floating reality" in some mysterious chaos.[6] If we feel suspended over a void, we have no secure footing enabling us to exist, much less to take action.

The actual experience of place means we are not adrift, not lost in a featureless void of infinite time or space. There are no substitutes for physical place, not even today's enticing location of a personal blog in cyberspace. Electronics don't pass muster. Nevertheless, a robust virtual reality, like the simulation where we pass through the black hole into Einstein's wormhole to find "the other side of infinity," can be a guidebook to finding place amid the dross. But virtual reality is not the real thing.[7]

Individual place is dense with evidence, our elementary surround, touching us far more directly than any symbol, cultural value, or social behavior. Meinig adds that "an individual piece in the infinitely varied mosaic of the earth" is where "all human events take place, all problems are anchored in place, and ultimately can only be understood in such terms."[8] This anchor has an absolute quality to it. Early twentieth-century philosopher John Dewey called this connection between self and place the "experiment of living."[9] Place is rich and profound as nowhere else. The philosopher Edward S. Casey seeks *kinesthesia*, the inner experience of the occupation of space that is more—a "complex qualitative whole"—than quantitative site or abstract space.[10]

The novelist and screenwriter James Agee grounded his life at his childhood home in Knoxville, Tennessee. A palpable reality emerges in an autobiographical fragment he wrote about the summer of 1915:

It has become that time of evening when people sit on their porches, rocking gently and talking gently and watching the street and standing up into their sphere of possessions of the trees, of birds' hung havens. . . . Now is the night one blue dew, my father has drained, he has coiled the hose. . . . The dry and exalted noise of the locusts from all the air at once enchants my eardrums. On the rough wet grass of the back yard my father and mother have spread quilts. We all lay there, my mother, my father, my uncle, my aunt, and I too am lying there. They are not talking much, and the talk is quiet, of nothing in particular, of nothing at all in particular, of nothing at all. The stars are wide and alive, they seem each like a smile of great sweetness, and they seem very near. . . . By some chance here [my family] are, all on this earth, and who shall ever tell the sorrow of being on this earth, lying, on quilts, on the grass, in a summer evening, among the sounds of the night. . . . Those receive me, who quietly treat me, as one familiar and well-beloved in that home, but will not, oh, will not, not now, not ever; but will not ever tell me who I am.[11]

Agee knew he alone would find out who he was on that same lawn. He was absolutely there, and nowhere else. That lawn was the bedrock of his existence. It enveloped him. Reflecting on the vivid memory, Agee knew that we are decisively objectified by what the place-world gives us.[12] We see the street corner, the mailman comes, the backyard sounds with play, weighty arguments linger, loved ones murmur. The outside world is noisy with voices or machinery, there are strange creaks or hums, or it is so silent that no-noise becomes loud. Ordinary place has a sacramental power in which we delight.[13] One of the best examples of this power is the unfolding of a single square mile of eastern Massachusetts by John Hanson Mitchell. He looked beyond private property, industrial development, and the spread of Boston suburbia to a deeply rooted geological and biological zone. Here Native Americans, in a mesmerizing dance, could still conjure up the momentary appearance of a sacred bear.[14] There must be more to place than what meets the gleam in the realtor's eye.

In Casey's words: "To be absolutely here means that *with my body* I am *in this place*: the very place my body stands or sits or walks in. To be here in this way is absolute in that it is not dependent on any 'theres.' . . . The absoluteness of my stand resists dissolution. . . . It affirms the uniqueness of the place I am in."[15] Casey adds, "Place subtends and enfolds us, lying perpetually under and around us."[16] In a vivid analogy he compares this personal connection to place with the Japanese art of origami, in which paper flowers exfoliate in water. Our place-world "opens up . . . making room for something more to take place." Such a place contains "intimate immensity and vastness," as if the origami paper, once wetted, will continue to open indefinitely.[17] Our anchored place is the location of current opportunity, offering multiple futures that are grounded in the present condition and a single past. It is a personal commons.

Irreducible concrete place holds true before we inhabit it, during our habitation, after our habitation, and whether we inhabit it or not. It is not vacant space. It is *before* anything else. Our place has the prospect of having been there long before we arrived, and it will remain after we are gone. We can trust it. It is original. It is perpetually fresh. It has, as the saying goes, an "authentic feel," like the family homestead or old summer cabin, the childhood neighborhood or schoolyard, the friendly workplace or comfortable home site. Aldo Leopold defined "the base datum of normalcy" in terms of the enduring ecological features of his revived farm homestead.[18]

In a striking parallel to Casey's exfoliating origami, the German philosopher Martin Heidegger pointed to the opening of the Open, the particular place—the Clearing—that makes room for the incarnation of our truest Self.[19] The world of our place cannot be dissolved into some dark abyss of empty, infinite space. Others link it to ownership of a piece of land as highly individualized private property. Place is already irreducibly there, inalienable, just as Thomas Jefferson inserted "inalienable rights" into the Declaration of Independence.

Americans, however, demand even more. We dream of utopias, be they rural, urban, suburban, or theme parks. Boston started as the "city set upon a hill," for the rest of the world to emulate. Upstate New York's "Burned-Over District" spawned scores of utopias—places of beauty and virtue. The Mormon search—"this is the place"—took decades. California is still alluring. Frederick Jackson Turner's 1893 notion of a continuous frontier is akin to Romanian philosopher Mircea Eliade's emphasis on reenactments of creation; Americans have repeatedly built a "First Place" by clearing trees and raising barns.[20] Whatever else, America has long been Nature's Nation. Cultural historian Perry Miller saw this in 1967 as he explored how the New England Puritans' attempt to capture the wilderness was instead taken over by America's geography, how the Transcendentalists of the 1830s insisted on a correspondence between Americans and wild nature.[21] The philosopher of geography Yi-Fu Tuan takes the next step. He celebrates the absolute qualities of nature that are beyond our understanding: "Doesn't nature provide even more powerful images [than our best myths]? What gives a better sense of calm than the sea at rest, or of exuberant energy than the primeval forest, or of vastness than the endless sweep of the plains?"[22] We thus move beyond the conventional myths of West or wilderness.

During World War II American soldiers, fighting in the European or the Pacific theater of war, spoke of their vivid memories of home place. Under life-threatening conditions in a strange place a sense of chaos and abandonment swept over them. The battlefield was not their habitation. Their capacity to hold within themselves an emotionally secure place, then, helped keep them psychologically stable under the stress of battle. They remembered their home's layout and the use of its rooms, as well as the artifacts that shaped daily life. Like Agee evoking his summer evening they held onto simple acts, even clichéd: getting dressed in the morning light of a second-floor bedroom, going to work in a steel mill, reshingling a roof, soaking out aches and pains in a hot bathtub, a family meal around the dinner table. They remembered a freshly plowed field, hunting with Dad in a patch of forest, a family picnic in the neighborhood park.[23] The "ordinary" place becomes the "great" place.

Place is a palpable reality, dense with evidence, our elementary surround, touching us differently than any symbol, cultural value, or social behavior. We are oriented by our physical place as by no other orientation. We cannot speak of human activity except as what occurred at a particular time in a particular place to particular people. This anchoring is not only private but public, our spaces including town squares, factory zones, and neighborhoods. We interweave our individual histories with those of our community and nation. Such diverse figures as the geographer Donald Meinig, the philosopher Martin Heidegger, and the landscape historian J. B. Jackson have written of how the labeling of public places shapes a nation's people.

The soldiers of World War II dug yet deeper to discover that they also shared public places with common characteristics: the home set amid city blocks, suburban cul-de-sacs, farmyards, and rural villages. The overseas soldiers were thus able to enjoy familiarity with their buddies: the New England village; the Main Street of Middle America; Iowa cornfields and California fruit orchards; Brooklyn; industrial cities like Pittsburgh, Cleveland, Gary, and Chicago. These places rang true. Brooklyn or Iowa became fixed in their minds as an American emblem of the good life. These places where they had dwelled in common, with shared beliefs and values, hopes and dreams, remained concrete, reliable, and enduring, while the rest of their world dissolved into life-threatening chaos. An implaced America is Authentic America.

Searching for Place in Third Nature

To the nonhuman natural world—First Nature—and our built environment—Second Nature—we have added another external world, Third Nature—the tantalizing power of virtual reality. A technologically adept person, skilled in negotiating virtual reality, may conclude that he or she is not placeless when inhabiting it but perhaps superiorly placed. The mundane world of ordinary life becomes lightweight, relegated to background noise. In his extraordinary 1977 book *Space and Place* Yi-Fu Tuan suggests the ease with which we continuously create and re-create new virtual realities: "With the destruction of one 'center of the world,' another can be built next to it, or in another location altogether, and it in turn becomes the 'center of the world.'"[24] One punch of the start key takes us into another spin in cyberspace. Virtual realities are the vivid outcomes of modernity. Living in our modern world, says Tuan, depends less and less on material objects and the physical environment.[25] Virtual reality is felicitous in contrast to the indifferent, even intransigent, external world. Invested with a fullness of imagination, virtual reality has no limits. If one world is unsatisfactory, then numberless others are available. It offers more opportunities for personal satisfaction than anything in the actual world.

As virtual reality becomes increasingly attractive as our primary home, our traditional ties to physical geography slip away from us ever more rapidly. We transcend the limitations of the ordinary world. Casey writes: "In inhabiting a virtual place, I have the distinct impression that the persons with whom I am communicating . . . though not physically present, nevertheless present themselves to me in a quasi face-to-face interaction. They are accessible to me and I to them. I seem to share the 'same space' with others who are in fact stationed elsewhere on the planet."[26] British geographers Martin Dodge and Rob Kitchin note: "One of the principal effects of cyberspace is the formation of communities that are free of the constraints of place and are based upon new modes of interaction and new forms of social relationships. Instead of being founded on geographic propinquity, these communities are grounded in communicative practice."[27]

The physical world is no longer our "living tether." The wonders of cyberspace and our daily ganglia of virtual realities bring into question the features and qualities of solid "place." Does it have so much finality? Cyberspace becomes a trackless region of possibility, as the American West became a mythic frontier of endless opportunity.[28] If we can no longer escape modernity's virtual worlds, where every place is far more a state of mind than a geographical location, then let's get better at inhabiting them. Educator David Sobel writes, "If our fingers continually just float above the keyboard, our minds will similarly just drift across the surface, never settling down, never developing a sense of place."[29] Physical places fade into oblivion, like the faces in an old photograph. To pick up Milan Kundera's phrase, we experience a lightness, transparency, and gradual disappearance of our actual being.

But inhabiting today's cyberplace doesn't work as an alternative. Virtual realities in cyberspace can only enjoy the formal qualities of geographic space. The best simulations are self-regulating and include personal immersion, but they are still simulations. The features of cyberspace are the ultimate deconstructions. Cyberspace is entirely socially produced, its form and structure dematerialized. Tuan anticipated the debate: "What can be known is a reality that is a construct of experience, a creation of feeling and thought."[30] The virtual realities that float in cyberspace are particularly rich constructs. The opportunities cyberspace offers—to create and re-create new spaces, new conditions, and new personalities—are ultimately its downfall. There is no concrete and intimate habitation of physicality. Dodge and Kitchin conclude that "the absolute geographical location of the infrastructure [of a Web site] is not important [i.e., relevant]."[31] Fabricated artificiality always remains as an undercurrent that reveals how false virtual reality is. The artifacts of games and simulations—indeed all virtual realities—have no weight or mass. While one of the virtues of games is repeatability, they concurrently leave no trace of their existence. They are more skittish than the elusive particles in the old physics cloud chamber. A simulation can defeat the very meaning of the word, becoming a simulation of itself and not of anything outside itself. Each Web site and each game is in fact its own map. More than mere simulations, they are self-referential, which is both their glory and their failure.

Heading Home

I have another bad habit, besides viewing the world through a camera's lens. I interview new acquaintances, and even strangers, as if I were a newspaper reporter. Whenever I'm in conversation with a group of people at a party, at a coffee break at a conference, or while waiting for the plane at the airport, I ask them where "home" is. Some will say, "Home is where I live now," but many will reflect for a moment, smile, and tell me about their childhood in Yonkers, Youngstown, or Yellow Springs. Their identification with "home" goes back, like Agee's, to first memories of a backyard, bike riding, or ice cream at a neighbor's. The memory of this "first place" is vivid, perennially fresh, and determinate.

Over the last dozen years I have troubled my students in environmental history and literature by asking them to put down on paper a mental map of their home place as youngsters of about ten years old. I tell them that one of a person's strongest assets is a highly individualized sense of place, but this doesn't mean much in the abstract—words in the ether. Their task is to map their own place as an embodied experience—their flesh and bones that inhabited bricks and concrete, windows and doors, greenery and open spaces. They are to map out their neighborhood at a time when their mobility was based on walking, riding a bicycle, riding in their parents' car, or taking the bus or train. Seasons, weather, even time of day are important features. Their highly personal map should include home, familiar routes to school, stores, places of worship, places of sports and recreation, public buildings—and the locations of relatives, friends, and jobs. The map is to show both human infrastructure and natural physical surrounds such as waterways, hills, parks, and other terrain. They are asked to identify friendly and dangerous places—the street corner where friends met, as well as the place where the dog bit them. Secret places might be included, where they could hide with best friends or even alone, to reflect on

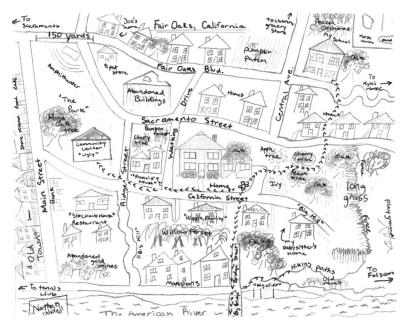

29. A student's cognitive map of Fair Oaks, California, ca. 2001.

their unobserved life. The handout for the assignment states: "Consider social activity, commercial appeal, personal places, handsome and ugly sites, indoor and outdoor activity. Name the school, store, park, relatives' or friends houses, etc. The map should describe crucial 'moments' in your life that would be informative, entertaining, and provocative for family and friends, and a real archive to share with your significant other(s) as well as your children and grandchildren."

Their neighborhoods' resonance is based on more than houses, intersections, and stores. These are their neighborhoods as they are dreaming them some ten to fifteen years later. It's true, in the words of Southern novelist Thomas Wolfe, that you can't go home again, because home, by your absence, becomes different from what it was when you left. One student from Arkansas was delighted, on returning home, to find that

little had changed. Another student, from the Bronx, had difficulty finding the old neighborhood he remembered, was deeply
disappointed, and found himself disoriented about his future.
What once was utterly familiar recedes into invisibility, becomes
a ghost landscape. Places look familiar and yet, when closely
observed, become strange. They become, paradoxically, nov-
elties. Looking back at home, the students feel like strangers,
even ghosts.

Most of my students, from a total of about ninety, are from
urban or suburban neighborhoods. Interestingly enough, most
of their cognitive maps cover approximately a square mile. But
size is not as relevant as the degree of intimacy and intensity
manifested. The rare student from a farmstead had a life centered on the farmyard, with long routes stringing off the map to
school, church, shopping, sports, jobs, and friends. One student
from mainland China found her life centered upon her housing compound. A student from Lebanon told of his mountain
village that had been dominated by four interlocked families
for seven hundred years. Some students, such as military kids or
the children of moveable corporate executives, have difficulty
identifying a single definitive home place because they had to
learn to inhabit a new place every two or three years. Yet they
discover individual local sites that are the center of powerful,
shaping, and lasting memories.

Memory and revisiting bring back iconic realities. California
philosopher Charles S. Peirce describes authentic childhood experience as a three-sided phenomenon: the reality of a child's
sandbox, his use of the word *sandbox*, and his mental picture
of a sandbox. Alone the sandbox cannot be seen "out there,"
while alone the word *sandbox* is merely a sound in the air; nor is
there any reality to a child's mental picture of a sandbox without the sandbox "out there." As a child Helen Keller was "blind,
dumb, and deaf" until the moment when she felt water flowing
over her hand from the pump and simultaneously realized that

the word *water*, spelled out on her other hand, referred to the same thing.[32] She had broken through the veil between self and the outer world.

In most cases students remember a "first place," the dominant matrix for their habitual actions. We can identify this place both as preobjective and as holding primal power. In childhood, says Casey, we are "plunged willy-nilly into a diverse (and sometimes frightening) array of places."[33] My students are surprised and impressed with the power their old (original) neighborhood still has over them and with how it continues to shape their expectations for the future. Work on their maps produces flurries of mental details. Mapmaking becomes an archetypal moment, focused both on the places recovered and on the realization of the impact of "first place." Anglo-American poet T. S. Eliot wrote:

> We shall not cease from exploration
> And the end of all our exploring
> Will be to arrive where we started
> And know the place for the first time[34]

Through the cognitive mapping assignment we learn the power of "ordinary places" when compared to the so-called great places such as Yosemite or Washington DC. The students discover that their home neighborhood is not indifferent or trivial, arbitrary or chaotic, but consistent, specific, and finely wrought. They as college students could not exist as ahistorical and antiplace beings. Place is not a corpse, but contains living peculiarities that generate thoughts and feelings and actions. Every place, no matter how dreary or mundane, contains the seeds of defining moments.[35] The students discover a secure place from which to look out at the rest of the world. In their mapmaking they deepen their sense of personal identity. It is an act of discovery (and creation) that is prior to any question of myth or reality. And my students are hardly alone in consistently identifying a paradigmatic childhood place. The Native American novelist Leslie Marmon Silko writes, "I was never afraid," because "the land all around me was teeming with creatures that were related to human beings and to me. . . . [I felt] a feeling of familiarity and warmth for the mesas and hills and borders where the incidents or action in the stories had taken place."[36]

My students discover, as George Washington Cable did in New Orleans, that their town "is a town that talks to you. The sidewalks talk to you, so do the shapes of the buildings. Everything here's telling you a story if you know how to listen, if you want that story. Some don't." Cable adds, "For me this city [New Orleans] is a mirror of the past as well as the present, a mirror in which I might one day see myself."[37] Another Southern writer, Eudora Welty, refused the temptation to see home as an abstraction: "Place isn't just history or philosophy; it's a sensory thing of sights and smells and seasons and earth and water and sky as well."[38] Literary historian Frederick Turner, writing of William Faulkner, agrees: "Here emerged a world, intact, solid, complex with its landmarks, seasons, smells, dialects, and customs. . . . It was founded in the soil of a real place. . . . It was his to work with as he would."[39] The playwright Sherwood Anderson assured a struggling Faulkner: "You're a country boy; all you know is that little patch up there in Mississippi where you started from. But that's all right too. It's America too." And Faulkner finally understood: "I discovered that my own little postage stamp of native soil was worth writing about and that I would never live long enough to exhaust it."[40] This realization resulted in his 1936 novel *Absalom, Absalom* and his 1942 short story "The Bear," works in which he received grace by coming to terms with his geography.

No one comes to mind here more powerfully than Mark Twain, who immersed himself in the small Mississippi river town of Hannibal, Missouri, in his near-autobiographical tales of Tom Sawyer and Huck Finn. Twain, like some of my students, was suspicious of this process—for two reasons. First, he felt a native distrust of "sham sentimentality." He raged against "the rot that deals with the 'happy days of yore,' the 'sweet yet melancholy past,' with its 'blighted hopes' and its 'vanished dreams'—and all that sort of drivel." Nevertheless, he was drawn back to Hannibal again and again, even though "the past can't be restored."[41] Second, as Frederick Turner notes, Twain was the representative placeless American, "in the midst of a ceaseless journey taking him ever farther from his origins."[42] Indeed, placelessness seemed to be Twain's trademark, considering his wanderings across the American West and his long tours of Europe. Nevertheless, as an adult he rerooted himself in Hartford, Connecticut, in contrast to the lack of concreteness that had characterized the first four years of his life, spent on a frontier farm settlement in Florida.

Home, however, was above all Hannibal. Determined to stay far away, Twain nevertheless was driven to return twice, in 1867 and in 1882, although he described the town as ordinary, even shabby, merely a place of dirty white buildings and fences needing whitewashing. Still, the town was a boy's paradise: the excitement, bustle, and commerce of the twice-daily steamboat arrival; the risky blandishments of rides on river rafts; fishing and exploring the shoreline; fighting fictitious "Injuns" in local woods; gawking at shoremen's eye-gouging battles; fearful ventures into a local limestone cave. The dark side that had kept him away for so long was characterized by his family's genteel poverty, especially after his father's death, along with an 1849 cholera epidemic; commonplace stabbings, shootings, and drunkenness; brutality toward slaves; the lynching of an unfortunate tramp; and the drowning of several friends.

Hannibal was also the home base for Twain's most memorable non-literary career, as a Mississippi riverboat pilot between

1857 and 1861, so well described in *Life on the Mississippi*. The river became its own extraordinary "place," this time defined by fluidity. For Twain, then, the river was also "home," an equally inexhaustible source of remembrance, growth, and personal identity. Twain put the river into a larger scale, calling it "The Body of the Nation."[43] It was "the great Mississippi, the majestic, the magnificent Mississippi, rolling its mile-wide tide along, shining in the sun; the dense forest away on the other side; the 'point' above the town, and the 'point' below, bounding the river-glimpse and turning it into a sort of sea, and withal a very still and brilliant and lonely one."[44] Twain truly never left the Mississippi, and the Mississippi never left him. It contained the delights of Paradise and the terrors of Hell—from the pleasures of every channel and bend, of idyllic steaming, and of the sovereignty of captainship, to the dread of boiler explosions and the challenge of shifting sandbars and tree-filled snags.[45] For Twain's sense of authenticity the river was more than equal to Hannibal.

In both worlds, of Hannibal and of the Mississippi, Twain found he had entered an authentic America that would feed the rest of his life, just as my students discovered as they constructed their childhood maps. For Twain Hannibal and the Mississippi were realizations of Whitman's "barbaric yawp," which frightened more genteel Americans (like Henry James) with its unkempt, brawling, violent conditions—regeneration through violence. Frederick Turner adds that Twain wondered whether Hannibal was the reality and the rest of his life a vapor: "It was as if he were awakening to the reality of his life from a long dream. *This* [both Hannibal and the Mississippi] was where he had lived, where he still lived in his mind."[46] He also feared that his experiences, as a child and as a riverboat pilot, had become irrelevant to modernity and that thus he too had become obsolete.

The life of Charles Ives overlapped that of Mark Twain. While

Ives made his fortune by reinventing the life-insurance industry at the turn of the twentieth century, he is better remembered today as America's *sui generis* classical-music composer.
Like Twain, and like my students, Charles Ives located his deepest origins—his music—in the physical places of his childhood
and youth, notably in Danbury, Connecticut, and Stockbridge,
Massachusetts. He wrote "Yankee Music." One cannot read the
titles of his music, or listen to it, without inhabiting the fresh,
vivid independence of a New England village.

Ives's youthful world in Danbury was idyllic, again like Agee's
summer Knoxville evening (which was set to music by Samuel
Barber in 1947–50). A Danbury poet wrote, "The whole of the
village swept round in a graceful curve line, thickly sprinkled
with buildings, and beyond, the tall . . . mountains made up
the backing of the picture in the most gorgeous array."[47] Ives,
backed by the security of home, dared to be openly experimental, deliberately naive, and playful, thumbing his nose at con-
vention. Already in 1891, at age seventeen, he transformed the
nation's unofficial national anthem, "America," into the scrappy
and musically blasphemous *Variations on "America,"* composing
on the local Congregational church organ, which he played
at Sunday services. (Later he would moonlight as organist for
New York City's Central Presbyterian Church.) As if combining
Europe and America, *Variations* contains parts written simultaneously in two keys.

Ives composed in a style drawn from the common human experiences of village childhood in the late nineteenth century
and the regionalism of Connecticut's New England, writing still-
unmatched authentically American music. As his biographer
Stuart Feder concludes, Ives offered "the music of patriotism
and religion; of history and politics; of family of self."[48] And particularly of identifiable local places. In 1917 Ives wrote evoca-
tive words to his song of childhood "The Things Our Fathers
Loved" (subtitled "And the Greatest of These Was Liberty")

I think there must be a place in the soul
All made of tunes, of tunes of long ago;
I hear the organ on the Main Street corner,
Aunt Sarah humming gospels;
Summer evenings,
The village cornet band playing in the square.
The town's Red White and Blue, all Red White and Blue
Now! Hear the songs! I know not what are the words
But they sing in my soul of the things our Fathers loved.[49]

Other favorite locations poured out in his songs: "The
Housatonic at Stockbridge" ("Contented river in thy dreamy
realm"); the opera house as the audience waits for the curtain
to rise ("The band is tuning up and soon will start to play");
even Ann Street, the shortest street in Manhattan ("Quaint
name, Ann Street. Width of same, ten feet"). Most famous remains Ives's *Three Places in New England* (1914, revised 1929),
which contains "The 'St. Gaudens' in Boston Common,"
"Putnam's Camp [Meeting], Redding, Connecticut," and "The
Housatonic at Stockbridge."[50]

Seasons and events that defined America took hold in Ives's
music, including Christmas, the Fourth of July, and Halloween.
Ives re-creates the annual summer visit of a traveling circus:

Down Main Street, comes the band,
Oh! "ain't it a grand
And glorious noise!"
Horses are prancing,
Knights advancing;
Helmets gleaming,
Pennants streaming.
Cleopatra's on her throne!
That golden hair is all her own.[51]

Remembering his astonishment as a four-year-old at the local revivalist camp meeting held just outside of Danbury, Ives reexperienced the blending of place and sound:

> I remember, when I was a boy—at the outdoor Camp Meeting services in Redding, all the farmers, their families and field hands, for miles around would come afoot or in their farm wagons. I remember how the great waves of sound used to come through the trees—when things like *Beulah land, Woodworth, Nearer My God to Thee, The shining Shore, Nettleton, In the Sweet Bye and Bye* and the like were sung by thousands of "let out" souls. . . . Father, who led the singing, sometimes with his cornet or his voice, sometimes with both voice and arms, and sometimes in the quieter hymns with a French horn or violin, would always encourage the people to sing their own way.[52]

Ives's Symphony No. 3 would be subtitled *The Camp Meeting*, its three movements labeled "Old Folks Gathering," "Children's Day," and "Communion." Other orchestral compositions include "Thanksgiving and Forefathers' Day" and the mysterious "Central Park in the Dark."[53] Ives described his *Holidays* symphony as "pictures of a boy's holidays in a country town"; it was written, according to Horatio Parker, his music professor at Yale, "hogging all the keys."[54] His Symphony No. 1, *New England Holidays*, includes the movements "Washington's Birthday," "Decoration Day," and "The Fourth of July," the latter two featuring marching bands sounding off as they cross each other's paths. Of the last movement, "The Fourth of July," Ives wrote:

> It's a boy's fourth—no historical orations—no patriotic grandiloquence by "grown ups"—no program in his yard! . . . Everybody knows what it's like—if everybody doesn't—Cannon on the Green, Village Band on Main Street, fire crackers, shanks missed on cornets [making them play "out of tune"]

... torpedoes [fireworks], Church bells, lost finger, fifes, clam chowder, a prize fight, drum corps, burnt shins, parades (in and out of step), saloons all closed (more drunks than usual), baseball game . . . pistols, mobbed umpire, Red, White and Blue, runaway horse,—and the day ends with the sky-rocket over the Church-steeple, just after the annual explosion sets the Town-hall on fire.[55]

Evil Places in the Sacred Land

Sometimes place can be as forbidding as a noose. In the sterile southern Appalachia country James Agee described the hard-scrabble life of three desperately poor sharecropper families. No one struggled more with, or found less reward from, their pieces of rented land: "The land that was under us lay down all around us and its continuance was enormous as if we were chips or matches floated, holding their own by their very minuteness, at a great distance out upon the surface of a tenderly laboring sea. The sky was even larger. . . . The sphere of power of a single human family and a mule is small; and within the limits of each of these small spheres [resides] the essential human frailty."[56]

Elsewhere, across much of urban America, the rewards of a consumer society were undermined by betrayal. The factories that built capital goods like electric generators and steam locomotives, or turned out consumer goods like refrigerators and washing machines, also pumped tons of black particles and noxious gases out of their smokestacks and into worker neighborhoods, threw scrap metal onto the empty lot next door, and dumped waste chemicals into the rivers. Industry has historically disposed its effluents in the "commons" of rivers, ponds, open fields, streets, and sewers, wherever there is a space in a city that no one owns or cares about—a hillside, streambed, backlot, curbside. Such pollution has long been dismissed as "the price of Progress," and in the process we have created uninhabitable ground.

In the 1970s land-use planner Ian McHarg wrote: "Epidemiologists speak of urban epidemics—heart and arterial disease, cancer, neuroses, psychoses. All of us record stress from sensory overload and negative hallucinations responding to urban anarchy. When you consider that New York may grow by fifteen hundred square miles of 'low-grade urban tissue'—[social critic] Lewis Mumford's phrase—in the next twenty years, you may recall [anthropologist] Loren Eiseley's image of our cities as gray, black, and brown blemishes on the green earth. These blemishes have dynamic tentacles extending from them. They may be evidence of a planetary disease—man."[57] McHarg compares the process of industrialization to a dangerous fungus or metastasizing cancer. It is a negative infrastructure, and recovery from it will be difficult and costly, if possible at all. Black holes—Superfund sites like Love Canal; nuclear-waste dumps like that at Hanford, Washington; industrial hog farms in North Carolina and Kansas; urban wastelands beneath highway interchanges—all create uninhabitable zones.

Mexican writer Kayta Mandoki finds that even sacred places are more than balanced by evil places: "Holy places radiate a great amount of energy to the pious. This grounds people, and individuals, not merely in a landscape, but in the power that creates and preserves the landscape." But she adds: "At the other extreme, Auschwitz and Treblinka collapse our capacity of understanding and faith in the same sense as black holes deplete all energy, even that of light. Just as Einstein tried to understand gravitation less as a force than as a distortion in the flat space-time continuum, so human geography also has its black holes that exert a tremendous gravitation field that absorbs all matter and energy. Whether a sacred place or a black hole, symbolically dense places tend to draw towards them further layers of meaning by warping their surroundings." Mandoki reminds us of the massacre at Amritsar in India on April 13, 1919; of the Spanish villagers of Guernica bombed by Nazi aircraft in April 1937; of the massacre of 33,771 Soviet Jews at Babi Yar in September 1941; and of the deaths of 300 unarmed Vietnamese villagers at My Lai in March 1968. We further remember almost 200 Indians massacred at Wounded Knee, South Dakota, on December 29, 1890.[58] The events of September 11, 2001, changed history, the day's disasters centered in identifiably American locations: New York City, the Pentagon, and a farm field in central Pennsylvania.

It's Hard to Stay Home, or Even Find It:
The American Conundrum of Place and Placelessness

On August 25, 1814, the British marched unopposed into Washington DC, U.S. troops having fled after destroying the Navy Yard. The British set fire to the Capitol, the White House, all of the department buildings, and several private homes, staying only a few days before a storm forced them to return to their ships. But something unexpected happened: in virtually all of history a nation has collapsed when its capital city is captured in warfare: Berlin, Paris, Rome, Richmond. We can reasonably conclude that the embattled Americans should have also collapsed.[59] But they didn't—not during the War of 1812 or, for that matter, during the American Revolution when the British, at one time or another, occupied the rebels' major cities, Boston, Philadelphia, and New York City.

If the heartbeat of the United States isn't found in its capital city, where is it? Is it in the Constitution? A careful reading of the Constitution reveals that the word *land* never appears, although political overlays like *state* and *territory* are present. Still, all in all, America's physical geography doesn't show up in the Constitution. Historian Daniel J. Boorstin claims that Americans enjoy a deliberate "vagueness" about their geography, a vagueness that grants them enough space—perhaps in the public domain, carved out of those outer places called "wilderness"—to encourage an open-ended democracy.[60]

The momentum of our mobile American society has historically been toward placelessness, or at least toward decisive interruptions of home place because of moves during childhood, education, job changes, and retirement. I was born in Chicago; lived in three houses in Riverside, Illinois; and studied in Greencastle, Indiana, in New York City's Manhattan Island, and back in Chicago. I then lived and worked in Urbana, Illinois; Pittsburgh, Pennsylvania; and Newark, New Jersey. My mother had arrived from Kolin in Bohemia to live in Cleveland, Chicago, and Fort Collins, my father from Lucke in Ukraine to live in Virginia, Iowa, and Illinois; my children, at least at this moment, live in Oregon, California, Ohio, Pennsylvania, Massachusetts, and Oeberusel, Germany. Where I now write these words, in a small community nestled in the Indiana Dunes, my connections are often less with the nearby fruit farms and tourist boutiques than with vegetables from Texas, fish from Icelandic waters, Internet assistance from India, coffee from Kenya or Columbia, and information from the *New York Times.*

An English traveler in the nineteenth century found little American loyalty to place: "There is as yet in New England and New York scarcely any such thing as local attachments—the love of a place because it is a man's own—because he has hewed it out of the wilderness, and made it what it is. Or because his father did so, and he and his family have been born and brought up, and spent their happy youthful days upon it. Speaking generally, every farm from Eastport in Maine to Buffalo on Lake Erie, is for sale."[61] A European visitor in 1847 wrote, "If God were suddenly to call the world to judgment, He would surprise two-thirds of the American population on the road like ants."[62] This placelessness is not surprising when we remember that a long-held America right is freedom of movement by deliberately autonomous individuals. Roots are inhibiting; we revel in rootlessness. Eric Hoffer, the poetic California longshoreman of

30. This popular Currier & Ives scene is mythical, a heroic landscape with Bierstadt-like sublime mountains in the distant haze. Although the wagon in the foreground, pulled by an unlikely six oxen, is struggling, its immediate landscape resembles a garden park, featuring lush grasses and deciduous trees (instead of mountain pines), as in the familiar East, and watered by a rushing stream. This was entry into the Promised Land. Even the two natives on horseback appear benign.

the 1960s, proudly proclaimed, "Only plants have roots, people have feet."[63] Yi-Fu Tuan adds that people also have imagination about places—mental pictures that they inhabit like virtual realities, that take them elsewhere faster than their feet can.

Westering Americans became like gypsies. A remarkably large number of people—about 350,000—trekked across the continent between 1841 and 1866. This image, of a one-sided movement across the continent, is compelling. Indiana essayist Scott Russell Sanders, no fan of the temporary, picks up historian Frederick Jackson Turner's frontier thesis: "Our Promised Land has always been over the next ridge or at the end of the trail, never under our feet. One hundred years after the official closing of the frontier, we have still not shaken off the romance of unlimited space. . . . In our national mythology, the worst fate is to be trapped on a farm, in a village, in the sticks, in some dead-end job or unglamorous marriage or played-out game. Stand still, we are warned, and you die."[64] The westward surge was an American instinct. Life is not a dead end after all.

The American migrants were unlike those who participated in the great centuries-long migrations of the past. They were not Genghis Kahn's hordes or marauders who settled down a generation or more in one place before moving on again. Instead the Americans crossed two-thirds of a continent in a single long odyssey that took several months. Overland emigrants made physically punishing, death-defying leaps across impossibly rugged spaces to reach the oases of California and Oregon. (This new freedom of movement, of course, belonged primarily to white males: slaves had as little mobility as prisoners, and women generally and wives especially were under the legal thumb of men.)

Then, between the Civil War and the 1920s, the greatest migration in world history—the movement of 40 million people from greater Europe to the United States—streamed mostly to crowded American cities. In addition, Southern blacks began their remarkable influx into Northern industrial cities—Pittsburgh, Detroit, Cleveland, Chicago—in the 1920s. American farm boys and girls continued to seek their fortunes in the city, to the extent that today less than 2 percent of Americans are still on the farm, compared to 30 percent in the 1930s. After World War II mobility came to define everyone's experience: the typical inhabitant of the model Long Island suburb, Levittown, moved once every two and a half years. In the 1950s, in Illinois's Park Forest, a third of the apartments and a fifth of the houses changed occupants yearly. Daniel Boorstin writes that this perpetual motion represented American society: "A small town was a place where a man settled. A suburb was a place to or from which a person moved.

Suburbanites expected their children to live elsewhere."[65] In most cases migration was from one metropolitan region to another: Pittsburgh to Houston, Chicago to Denver, Cleveland to Los Angeles. One out of five Westerners relocated annually in the 1990s. According to the 2000 census, of 280 million Americans a remarkable 40 million relocate every year.[66] In this view the True American carries his place with him, like the turtle his shell. Many Americans inhabit so-called mobile homes that are fixed in trailer parks, but their wheels and hitches still signal rootlessness. Mircea Eliade, who was born in Romania, lived in Paris, and made his home in Chicago, observed that the novelty of America was based on the idea of continuously re-created new home places.[67] The Indian-British novelist Salman Rushdie goes even further, celebrating "the migrant sensibility" as a "radically new type of human being" who inhabits a liberating virtual reality: "people who root themselves in ideas rather than places, in memories as much as in material things."[68]

Second Nature's victories of engineered infrastructure, inflated consumerism, and ungoverned triumphalism also place traps in the way of our steps toward place. The concreteness of personal space, and hence our attachment to it, too easily slips beneath the surface of the world's noise—the fragmenting force of sensationalist media, the opiate of sports, the solvents of globalization. Great distances are folded into instantaneous connections. American society now tends toward interchangeable landscapes. Philosopher Deborah Tall, for example, writes, "Like so many Americans raised in suburbs, I have never really belonged to an American landscape." Indoor shopping malls are made deliberately identical, down to their food courts. Tall quotes Gertrude Stein: "When you get there, there is no there, there." Tall concludes that "the fear of living nowhere in . . . no place, or in an indistinct clone town, is a very real one."[69] The problem arises when we cannot tell the difference between a supermall outside Houston and a supermall outside Albuquerque.

Once-uniquely identifiable landscapes have become a common hash. Downtown Atlanta is comparable to downtown Houston, which compares to downtown Denver; I can only determine my location by checking auto license plates. Newsstands may offer the local newspaper, but it is hidden behind USA *Today*. With the appearance of such solvents, human activity moves ever more easily across increasingly porous boundaries. Individual localities are more and more exposed to the vagaries of the flows of population, ideas, investments, goods, weapons, disease, invasive plants, and pollution.[70] Places can change beyond recognition within a generation.

Place, no matter how palpable, is a fleeting asset. Often the harder my students work on their mental maps of home, the more the reality of each place recedes into vague and piecemeal memory. Our search for utopia—that small perfect place that we can inhabit at exactly the right moment—is a fragile one indeed. Where we expect to feel most at home, we instead feel homeless—alienated.[71] No wonder the rush to controllable virtual realities, to man-made simulations in cyberspace. Landscape essayist John Brinkerhoff Jackson worries over whether a genuine American landscape still exists. Deborah Tall longs to abandon landscapes stripped of nature, devoted to homogeneity and indistinctness—instead "to feel the world as unavoidably real, even if ferocious."[72] (This reality and ferocity are what Annie Dillard found at Virginia's Tinker Creek in the 1970s.) Tuan reminds us that the reality of places marks them as "centers of felt value where biological needs, such as those for food, water, rest, and procreation, are satisfied."[73] We need to inhabit these palpable, dense, and elementary places. A street name and number in Tupelo, Mississippi, is vastly different from an identical one in Boise, Idaho. Strolling along Chicago's lakefront is different from strolling along Manhattan's West Side or San Diego's shoreline.

Our search is not a spectator sport. Americans adrift are

ghosts at a feast, denying the reality of personal place. The philosopher of travel Peter Matthiessen agrees about the difficulties: "The journey [to one's home place] is hard, for the secret place where we have always been is overgrown with thorns and thickets of 'ideas,' of fears and defenses, prejudices and repressions."[74] Yi-Fu Tuan argues that humans can for a brief time, only when they are infants, know how it feels to live in a nondualistic world where impressions and their sources are one and same.[75] Others suggest that we fatally enter our perceptual prisons certainly no later than age sixteen. This information may not take into account the power of media and games to accelerate our perceptual trap.

The English novelist Lawrence Durrell summed up placelessness: "For the modern self, all places are essentially the same: in the uniform homogeneous space of a Euclidean-Newtonian grid, all places are essentially interchangeable."[76] Humanity's quest for truth during most of the twentieth century was also complicated by the acknowledgment of indeterminacy: the very presence of the observer, and his actions, intrudes in even the most rigorous scientific experiment, shaping the results of the research. This was the complaint of the German physicist Werner Heisenberg, who warned, "Remember that what we observe is not nature in itself but nature exposed to our method of questioning."[77] We ask questions in a particular way, and the answers come in the same particular way—maybe or maybe not the set of questions and answers that can best uncover the exterior skin of reality, much less its innermost core. Nor does nature provide assistance, since it remains indifferent to our plight.

Environmental pioneer Paul Shepard worries that this surrender of real place points to "the final surrender to the anomie of meaninglessness." We "become identical atoms suspended in an infinite Newtonian void that is otherwise featureless." Placelessness makes us "regress through an ecological floor—the lack of place and time and the non-human contin-

uum—to farcical substitutes and . . . the madness of deprivation—the absence of sky, earth, seasons, nonhuman life, and
finally, [our] own identities as individuals and species, all necessary to our life as organisms."[78] In such placelessness there
are no lasting scenes to offer a plenitude of experience, re-
flection, and memory. This new condition lacks the stability
of traditional hands-on societies, where the physical features
of the immediate landscape largely determine how people live
and work. As in the Beatles song "Nowhere Man" placelessness
points us toward the void: "He's a real nowhere man, sitting in
his nowhere land." Nowhere Man is today's Everyman: "Isn't
he a bit like you and me?" The song further points to opportunity missed: "Nowhere man, the world is at your command."[79]
Instead, however, Everyman passively, even helplessly, inhabits
no place, but the void. Filmmaker Ingmar Bergman struggled
with the placelessness of modernity. It was all-encompassing:
"This world is a place where faith is tenuous; communication
elusive; and self-knowledge illusory at best."[80] This nowhereness
is even more affecting than the chaos of downtown Detroit or
Los Angeles. T. S. Eliot's *Wasteland* agonized in 1922, present-
ing haunting images of empty minds and empty lives cast adrift
in a placelessness that was profoundly resistant to the living link
between the human self and physical place. Landscape essayist
J. B. Jackson describes today's "existential landscape [that] is
already taking form around us." It is emptied of its own meaning, without absolutes, without prototypes, devoted to change
and mobility."[81]

Place itself has become plastic, stretched and shaped like
taffy. Our placelessness has been a long time coming and is
not simply the outcome of the recent cyberworld. It is, indeed,
the realization of Marshall McLuhan's 1960s "global village,"
its universal daily life taking place on the computer screen.
The French philosopher Michel Foucault, echoing Martin
Heidegger, observed that the twentieth century was above all

31. A detail of Claude Fiddler's *Bristlecone Pines* (2005).

the opening epoch of collapsed space and time. Foucault wrote: "We are in the epoch of simultaneity . . . juxtaposition . . . [of] the near and far, of the side-by-side, of the dispersed. . . . Our experience of the world is less that of a long life developing through time than that of a network that connects points and that intersects with its own skein."[82] The Czech political essayist Milan Kundera describes this condition as "the time of a humanity that has lost all continuity with humanity, of a humanity that no longer knows anything nor remembers anything, that lives in nameless cities with nameless streets or streets with names different from the ones they had yesterday, because a name means continuity with the past and people without a past are people without a name."[83] The very elements of place are deliberately sheared off: physical identity, local character, and a unique history—the "presence" of place.

The Authenticity of Nature: *One Man's Bristlecones*

Back in 1958 the dendrochronologist (tree-ring scientist) Edmund Schulman, writing in *National Geographic*, told the world that the bristlecone pine of America's desolate Great Basin was the world's oldest known living thing.[84] But it was Michael P. Cohen, biologist and literary historian, who found in this rugged and dramatic tree an icon of physical place.[85] As Cohen's pickup rattled down the dirt road of White Mountain in California's eastern Sierras, twelve miles past the Schulman Grove, to the Patriarch Grove at 11,000 feet, he discovered a "moonscape." This place is, according to visitors, "a place where life is a miracle." Cohen discovered that "here a single [iconic] bristlecone—a remarkable tree—seems to call attention to timberline, to its own dilemma, caught at the edge of possibility."[86] Here there are limits, edges, and frontiers. Here place is defined by cold, by wind, by desiccation, by soil chemistry, by rarefied atmosphere, and even by intense light. An American wildlands dreamworld, the bristlecone's territory is unfriendly to humans, who can only temporarily visit the arid zones of the Great Basin, noted for its extreme climates, especially at 9,500 to 11,650 feet. This is not easily accessible First Nature; it is kin to Burke's sublime. The flamboyant explorer who named the Great Basin in 1844, John Charles Fremont, simply described it as "unnatural." Cohen clarifies this view: "The Great Basin's natural history is filled with things not like us, things which seem out of the past, as frightening as eternity, suggesting time out of mind."[87] The bristlecone at its own edge of survivability points to apocalyptic possibility.

Our short American history seems trivial compared to that of a single species (*Pinus longaeva* D. K. Bailey), which has endured for as many as five thousand years. The tree has become a physical symbol of the metaphysical vistas of the American Southwest. It lived in America long before Europeans arrived; with any luck, by avoiding human intervention, it will be around long after we are gone. The Polish émigré poet Joseph Brodsky is even more emphatic about the importance of the autonomy of First Nature:

There is a difference between the way a European perceives nature and the way an American does. . . . When a European . . . encounters a tree, it's a tree made familiar by history, to which it's been a witness. This or that king sat underneath it, laying down this or that law—something of that sort. A tree stands there rustling, as it were, with allusions. Pleased and somewhat pensive, our man, refreshed but unchanged by that encounter, returns to his inn or cottage . . . and proceeds to have a good merry time. Whereas when an American walks out of his house and encounters a tree it is a meeting of equals. Man and tree face each other in their respective primal power, free of references: neither has a past, and as to whose future is greater, it is a toss-up. Basically, it's epidermis meeting bark. Our man returns to his cabin in a state of bewilderment, to say the least, if not in actual shock or terror.[88]

Michael Cohen's bristlecone refuses to be enclosed by our views; it has its own identity and history that, as Brodsky suggests, has nothing to do with us. Yet Cohen adds that "the bristlecone has become a vehicle which holds a flow of human meanings attached to it."[89] It embodies Nature in America. For decades the Forest Service called the bristlecone useless, until the tree's age radically changed its mind. The scientist Dana K. Bailey, for whom the tree was named, waxed poetic: "What is remarkable about the ancient trees is the sense of serenity and peace in the face of great environmental hostility, and the scars of a very long life in the hostile environment, which they impart."[90] Described in human terms, these scars make the trees grotesque, deformed, distorted, gnarled, and weird. That leaves us, in Brodsky's words, in a state of bewilderment, shock, and terror. Michael Cohen continues: "Certain trees, nearly five thousand years old, have stories to tell. . . . This is no dream. In adjusting their sense to the trees and what they have been told about them, people confront a dream and a reality . . . [including] a confusion, doubt, or agitation of mind." Cohen adds that the age of the trees points toward the infinite and eternal, "beyond human sentience, freed from human subjectivity."[91]

Like a radar beam we "lock onto" the bristlecone so that it becomes a single entry point into First Nature. We set aside Second Nature—our built environment—as lacking the primeval power of place that we long for. The bristlecone is a scraggly tree, with dense clusters of dark needles, likened to the bristles on a bottle brush. Cohen says that the trees "seem all elbows," contorted, stressed, sometimes shrubby, sometimes spike-tipped, mostly fifteen feet tall but sometimes reaching fifty feet. The bark is light brown to reddish-brown. "They are," he adds, "not crooked so much as they are bowed or bent, seeming to submit or yield to the environment, but resisting it too. . . . These trees, as bodies, have complex articulations with their environments."[92] Cohen identifies the features of authenticity that belong to the trees: adversity, eternity, antiquity, life-and-death struggle, desolation, solitude, the grotesque, longevity, stubbornness, survival, and nostalgia. Yet these are human words to fill the trees' own silence.[93]

What is gnarled and weird is also beautiful because it is old, almost like stone, proof of longevity under adversity. In August 1964 a worker for the Forest Service was ordered by the district ranger to cut down a bristlecone on the Wheeler Peak moraine in Nevada, near the Utah border. It was needed for scientific research. But according to an eyewitness, Mike Drakulich refused to start up his chainsaw: "He walked over to the tree and touched it. 'I'm not cutting this tree,' he said."[94] Drakulich was no effete Eastern aesthete, but a hard-working local inhabitant. He "apparently was unable to do the act . . . that required hardening the human heart." Since the tree was cut down the next day by other men, Cohen concludes that Drakulich's "was no small act of mutiny. It was a powerful, if symbolic gesture." Cohen quotes aesthetician Terry Eagleton

32. Valerie Cohen's watercolor painting *Bristlecone Pines* (1998).

to explain Drakulich's action, citing "the body's long inarticulate rebellion against the tyranny of the theoretical"—including the assignment to cut down an ancient tree for the sake of science.[95] Beauty is tied to place by physical contact and careful looking. Other visitors to the moraine had been so affected by the bristlecone that they named it the Prometheus Tree. It was, they said, exceptional, giving off a distinctive "presence," confronting visitors with the power of the eternal.

Joseph Brodsky concludes that we Americans still inhabit an untheoretical, ahistorical world—"present at the creation," epidermis rubbing against bark. Cohen adds, "Handling the trees, you are constantly reminded of contrasts between their concrete aspects: the substantial, resilient, and sturdy; the ethereal, fragile, and delicate; rough and smooth; accessible feelings and hidden ones." Trees are substantial: "They are striking material objects, and that is what allows them to be made into aesthetic representations. . . . There is nothing subjective about these [climate] conditions or the trees shaped by them. There is something subjective in the way people represent them." Cohen concludes that aesthetic responses to bristlecones may be subversive, because trees are capable of changing the perceiver's sense of his or her own body.[96]

Cohen reminds us that "these trees lived in these regions before humans conceptualized them." For Cohen the bristlecone pine becomes like a thought-provoking book, "offering an unending dialogue between the reader and the text."[97] He follows the lead of Italian philosopher-novelist Umberto Eco, who wrote of using a text or a tree for daydreaming, which is "not a public affair," says Eco, for "it leads us to move within the narrative wood as if it were our own private garden."[98] Brodsky describes the European tree as rustling with historical allusions; for Americans the tree, says Cohen, is "a brute reality that is solitary by nature, fragmented, asocial, unable to enter into relationships . . . set free from and broken off from values of the past." He adds, "When the individual is isolated for a long time, eternity seems to draw near."[99] Cohen notes that "observers keep going back to them [the trees], hoping to ask the right questions and finally understand something about change and longevity—ultimate human questions."[100]

Tree-ring scientist Valmore C. LaMarche Jr. writes: "The lifespace of a single ancient tree encompasses the whole period of the development of our urban, technological western culture. Can we build anything that we can be sure will survive for 5,000 years?"[101] The irony of the Great Basin, says Cohen, is that it was also the location of numerous nuclear weapons tests. It is further the site of America's ultimate metropolis, our fastest-growing city, Las Vegas. Cohen adds, "The landscape [of Second Nature] is saturated with human acts which seem terrible, and not like us. Even the people I meet often seem damaged, fragile, or strange, and I worry about them."[102]

Michael Cohen sought the bristlecone heights as if he were

MAP 1. U.S. Geological Survey Map of U.S. Counties (2005)

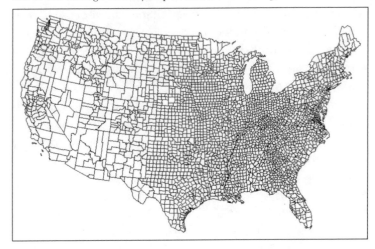

1 and 2. Counties have historically been the fundamental U.S. units of politics. Their boundaries, mostly geometric, do not reflect natural phenomenon, but their data is applied to environmental conclusions. One viable alternative focuses on watersheds, or hydrological unit boundaries, often mini-ecosystems. The map of the United States changes dramatically when shaped by watersheds rather than counties. Such a view trades one set of perceptions for another. Philosopher of science Thomas Kuhn might have labeled the difference a "paradigm shift."

a pilgrim: "Perhaps this has been a way to hide from modern realities too frightening to confront on a daily basis, but it has also exposed me to another reality . . . [that of] the bristlecone pines."[103] He notes especially the tree's "physical form, which insinuates, even if it doesn't answer, questions of being and knowing, of life and death, beginning and end."[104] Great Basin essayist Darwin Lambert writes, "We walked up to it and touched it, caressing the twisting, flowing grain of its naked wood . . . encircling its trunk with our four arms and our bodies . . . on a deeper level perhaps to embrace it as a fellow member of the lost community of life."[105] A Forest Service handbook is unusually eloquent: "There is a definite emotional impact upon meeting a 4,000 year old tree."[106] Utah essayist Terry Tempest

MAP 2. Hydrological Unit Boundaries (1998)

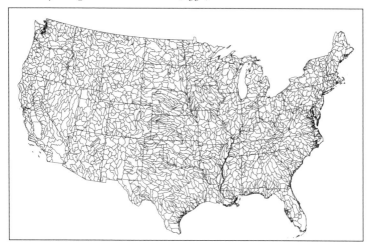

Source: U.S. Department of Agriculture, Natural Resources Conservation Service, Resource Assessment and Strategic Planning Division, Washington DC, July 1998. Map ID: 3862. For proper interpretation, see Explanation of Analysis for this map at USDA website (search for "USDASOTL" to locate map index).

Williams also tells a story: "A friend the other day told me a story of walking up to a particular ridge where a bristlecone pine stood, one of the oldest trees on Earth. He considered it his Elder and went to pay his respects as he had done year after year. When he finally found his way to the tree, it had been cut down. The body of the bristlecone pine lay on its side sawed into pieces. He stood before the stump for some time and then pulled out his pocketknife and made a small cut along the tip of his thumb. He let the blood drip onto the stump."[107] Brodsky's epidermis meets Cohen's bark.

The Way to Save Nature from Its Competitors: *The Drama of Ecosystems Unveils the Power, Complexity, and Beauty of First Nature*

Economists have long spoken of the "invisible hand" of the marketplace in shaping material life. The natural world, in contrast,

MAP 3. Major Land Resource Area (MLRA) Boundaries (1998)

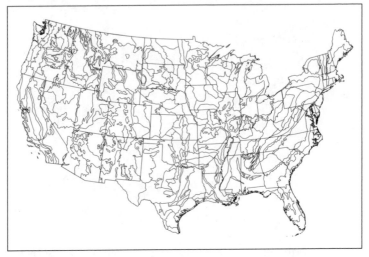

3. The boundaries on this map ignore county lines and other political borders to define geographical regions according to their agricultural usefulness. Not as extreme as boundaries defined by ecoregions (map 4) or as natural as boundaries defined by watersheds (map 2), MLRA boundaries nevertheless indicate a major shift in understanding humanity's relationship to the natural world.

Source: U.S. Department of Agriculture, Natural Resources Conservation Service, Resource Assessment and Strategic Planning Division, Washington DC, July 1998. Map ID: 2147. For proper interpretation, see Explanation of Analysis for this map at USDA website (search for "USDASOTL" to locate map index).

is a powerful "visible hand" that only seems invisible because we misunderstand or ignore it. It is hidden behind the billboards, diminished as I drive on automatic pilot to the mall. Nature is not merely the disposable stage set for some larger human drama. Nor can it be measured by dollar value and markets alone. Natural geographic systems—ecoregions, biotic provinces, physiographic provinces, biomes, and ecosystems—exist beyond the reductionism of traditional science to interweave the sciences—geology, hydrology, soil science, biology, and atmospheric science—together with the social sciences and humanities—history, economics, psychology, philosophy, and literature—to create an expansive lens that we can use to view our

MAP 4. Ecoregions and Metropolitan Areas (2001)

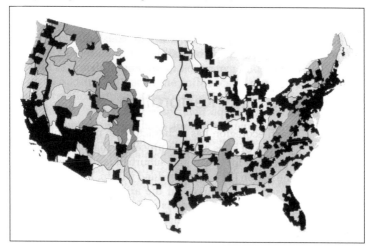

4. Between 1976 and 1998 Robert G. Bailey of the USDA Forest Service pioneered the identification of ecoregion boundaries using twenty principles, including climate zones; landforms such as mountains, grasslands, and deserts; pioneer vegetation; soils; water; and geology. Unlike in maps shaped by MLRA boundaries, human usefulness is not a factor; natural interdependency is crucial. This map displays Bailey's ecoregions, overlapped by major metropolitan regions (in black) that are continuously expanding into natural divisions. The implications of this view are just being explored, but the fact that metropolitan regions are blacked out here strongly suggests that ecoregion principles cannot apply to such areas. Bailey concludes that the viability of ecosystems depends upon their wider context, whether human development matches the limits of ecosystem regions.

world and our place in it. We discover a world that is fractal: no matter how we continuously dissect nature, the complexity remains.[108] We learn humility as the world remains beyond total understanding, much less total control. Bioregionalism based on ecosystems is the best template we have yet devised, a good match for the realities of our geography.

For 250 years we have surrendered to political boundaries, territories, states, counties, and the grid of the land system. Alternatives, however, are gradually emerging. Since the 1940s and 1950s Erwin Raisz's famous physiographic maps have shown state lines disappearing beneath geographical features, cities becoming mere dots on complex landscapes. By the 1980s federal agencies like the Natural Resources Conservation Service (formerly the Soil and Water Conservation Service), the Geological Survey, and the Environmental Protection Agency were measuring the nation's geography according to watersheds—zones where the movement of rivers and streams defines the landscape. A 1981 handbook by the U.S. Department of Agriculture, "Major Land Resource Area (MLRA) Boundaries," organized its maps according to conservation—highest efficiency—for human consumption. Regardless of state and county lines, the handbook's maps are divided according to land use (farms, ranches, dryland farming, irrigation farming), elevation and topography, climate (precipitation, temperature, freeze-free period), water (precipitation, rivers and streams, groundwater), soils (texture, moisture/temperature regimes), and potential natural vegetation (grasses, shrubs, trees).[109] In parallel with the MLRA agricultural criteria U.S. Geological Survey scientist Robert G. Bailey pioneered a strictly environmental approach that emphasizes natural soil, terrain, and climate conditions. Bailey concludes that if one begins with such ecosystems, they invariably cross jurisdictional, ownership, and state boundaries, making counties and sections far less relevant.[110] Today's dominant geography, he says, must be laid out as bioregions. Despite this impressive beginning, however, the shift to an ecological approach has come so recently that America is still mostly an unknown in environmental terms. Indeed, one can claim that no one knew what America's geography was like until ecologists began applying ideas of biomes, ecosystems, and interdependency to the landscape.

There are moves under way to shift the borders of Yellowstone National Park from its current squared-off lines to a larger, irregularly shaped entity—the Yellowstone Ecosystem, based on watersheds, topography, geological formations, and flora and

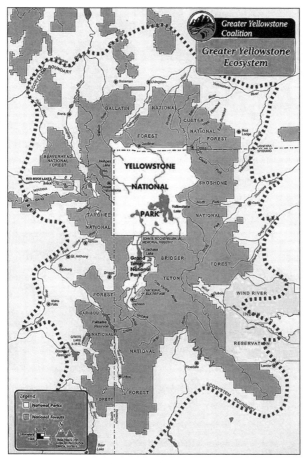

5. When Yellowstone was designated a national park by President U. S. Grant in 1872, its boundaries, enclosing 2.2 million acres, were drawn to include all the regional geyser (geothermal) basins. These were to be protected as "natural curiosities." An ecosystem approach surfaced a century later, first designated to include the range of the grizzly bear. The first informal minimum boundary lines of a Greater Yellowstone Ecosystem marked out 4 million acres. In 1994 this area was enlarged to 20 million acres, based on the range of grizzly bears, Yellowstone cutthroat trout, pronghorn antelope, beaver, ungulates like elk and wolves, and plants like whitebark pine and quaking aspen. Changing scientific knowledge, along with field data, has changed the perceived boundaries of wild places. This paradigm shift cannot be overemphasized; it has changed our national perception of America's geography.

fauna. In north central Nebraska the Nature Conservancy is seeking to define, recover, and restore the unique Niobrara River ecosystem. We now know that Florida's Everglades involves a water system that covers most of the southern half of the state, but it is fragmented in both public and private hands—agricultural, urban, and quasi-protected within a national park. Clearly, there are flaws in property laws and market processes. What we are beginning to learn is that the more the land is studied, the less divisible it seems to be.

Bioregionalism can be a middle ground, linking personal place and national history—the glue of what it means to be an American. Kevin Lynch points to 295 possible regions to which Americans see themselves as belonging.[111] I am learning not just to inhabit my beloved dunes but to be a citizen of the entire and international Great Lakes Region, with its distinctive and interlocked issues of fresh water, land use, native and invasive species, and urbanization. Daniel Kemmis, of a Western think tank, describes this shift as "learning to think like a region," such as the Pacific Northwest or the Rocky Mountain West.[112] Dan Flores urges us to fold the "tangible ecologies of place into human history. . . . We remain biological even with all our bewildering array of cultural dressings." Flores identifies the "close linkage between ecological locale and human culture. . . . In a variety of ways humans not only alter environments but also adapt to them."[113] Flores joins with ecologist Peter Berg in advocating "a kind of spiritual identification with a particular kind of country and its wild nature [that provides] the basis for the kind of land care the world so definitely needs."[114] With the appearance of an environmental viewpoint we are learning that America's nonhuman world is infinitely lavish and fertile, perpetually dynamic, a continuously moving point on a moving line. It is still a paradise in which we sleepwalk.

Nature doesn't reveal itself readily, nor does it conveniently release the information we need from it. Columbus did bump into an unexpected real place; his task was to properly identify it. Kansas's wheat farmers had to learn that their real climate is drought, not rainfall. Part of accepting nature is acknowledging its mystery. The Midwestern essayist Scott Russell Sanders writes, "The achievements of science delude many into thinking that we have graduated from nature, that we can only scorn conditions as we see fit, that we are the bosses of the universe." He adds:

> Nature is noisy, all right, and it moves in patterns, but neither sounds nor movements seem to be addressed to us. The racket the world makes is either pure babble—as in the crash of waves, the crackle of lightning, the sizzle of wind through trees, the static from stars—or else it is a language not meant for human ears, as in the calls of dolphins and owls. . . . Like refugees washed up on a foreign shore, we spy and eavesdrop on nature, searching for clues, trying to decipher an alien tongue. . . . Since nobody has supplied us with a cosmic dictionary, we have been laboring, word by word, over a thousand generations, to compile one for ourselves.[115]

The new environmental science comes closer to seeing "reality" than ever before. This frontier science has begun to peer directly—through a glass clearly—at America's actual landscape. This is a matter of proper correspondence to reality. The starting point determines the finish line.

Inhabiting One's Place: *The Resettlement of America*

Man wants his physical fulfillment first and foremost, since now, once and once alone, he is in the flesh and potent. For man, the vast marvel is to be alive. For man, as for flower and beast and bird, the supreme triumph is to be most vividly, most perfectly alive. . . . The dead may look after the afterwards. But the magnificent

here and now of life in the flesh is ours, and ours alone, and ours
only for a time. We ought to dance with rapture that we should be
alive and in the flesh, and part of the living, incarnate cosmos.

— D. H. LAWRENCE, 1929

Relearning the Story of Place in American History

Henry David Thoreau looked for "the wild" as he strolled in
Walden's environs. Literary historian Frederick Turner argues that Thoreau's quest stemmed from his "powerful telluric
drive." Turner adds that Thoreau taught all of us to dig deeper into aboriginal nature, to investigate "what lay beneath the
soil on which he walked, . . . what was beneath the furrowed
fields of his neighborhood . . . that 'hard bottom and rocks
in place.' . . . Was there not a lower layer yet to be explored, a
stratum that if probed might yield up the quintessential spirit
of the land?"[116] The most salient features of place come to light
when, with Thoreau, we allow place its "wild" identity. Our sight
is enlarged and intensified. We are raised to another power.
Turner, following in the footsteps of Willa Cather's *Death Comes
for the Archbishop*, found that in the landscape of central New
Mexico "the country suddenly got bigger." "In this new ampli-
tude," he notes, "I felt truly on my way." He describes entering
"a bladelike cleanness."[117] This enlarged landscape forces him
to confront the limits of his own humanity. The challenge here,
learned from Thoreau and Cather, is to remain open, naked
and unguarded, to one's landscape, to look anew at existence
and sensations, no matter how strange the outcome.[118]

We thus immerse ourselves in another dimension of habitation. Place becomes fundamental, more primeval and fresh.
We stretch the boundaries of our actual existence. We are em-
powered by personal engagement with our surroundings. As I
type this on my laptop, I am also beginning to learn to inhabit
my own home place, situated off the shore of Lake Michigan. I
work in my living room, dominated by its large stone fireplace.

Secure in my home place, I experience the limits of the metaphors of language and the quantifications of science. Yet I use both. I learn that we boast too much that we have conquered our geography. I am humbled by so little reliable knowledge of my territories.[119]

No name is more revered in contemporary environmental circles than Aldo Leopold's. No book is more treasured than *A Sand County Almanac*, published in 1949, after Leopold's death of a heart attack while fighting a brush fire. Leopold had scientific credibility as a technical forester and a pioneer in wildlife management, and near the end of his life he incorporated a carefully thought-out land ethic into his scientific worldview. His essay "The Land Ethic" is the undisputed sacred text of the modern environmental movement. In 1935 Leopold and his family began planting pine trees, as many as six thousand a year, together with other trees, shrubs, grasses, and flowers, on an abandoned eighty-acre farm. He turned the farmland, with its weekend chicken-shed shack, into a working laboratory in an effort to restore the worn-out land to its state of "aboriginal health." He had watched too many New Deal cleanup crews do more harm than good when they took out brush needed for wildlife food and shelter, silted trout streams, and planted rows of identical trees: "Land, then, is not merely soil; it is a fountain of energy flowing through a circuit of soils, plants, and animals. Food chains are the living channels that conduct energy upward; death and decay return it to the soil. The circuit is not closed. It is a sustained circuit, like a slowly augmented revolving fund of life."[120]

Leopold, using his experience as a private landowner, began shaping an alternative view of the rights and duties accompanying private property. In this case scientific knowledge requires an ethic that recognizes the right to existence of all components of the ecological chain: the real end is a *universal symbiosis with land*, economic and aesthetic, public and private. A land ethic means that privately owned land demands owner responsibility if it is to be restored to ecological integrity. Responsible landowners are rewarded with a powerful sense of belonging to something greater than themselves. According to Leopold, ethical duty toward the land in particular means restraint in its use, thus casting doubt upon the potential of development to raise its economic value. This is a radical and controversial concept. An obsession with profit, Leopold declared, not only leads to the destruction of the environment but also trivializes the more noble aspirations of human civilization.

Leopold brought up to date the alarm that Thoreau had felt about the flawed human-environment connection in a materialistic civilization. Leopold took earlier holistic but vague philosophies and turned them into science. He dreaded the romantic effusions of popular "nature fakirs," the badly informed protectors of wilderness, who had "a zeal so uncritical—so devoid of discrimination—that any nostrum is likely to be gulped up with a shout." While most of forestry and land management science had become "narrow as a clam," Leopold went in the opposite direction, turning to the new science of ecology. He was not without his critics. Scientists doubted his ecology and philosophers his ethical theories. But Leopold retorted by saying that scientific fact could do damage unless it was used responsibly, and that moral sense went begging unless it could be applied to a problem scientifically.

Leopold also expanded upon the long-standing American tradition that made scenic beauty an important criterion for successful conservation and preservation. He enlarged the idea of the beautiful beyond the classic mountains, waterfalls, canyons, and lakes to include nonscenic swamps, dunes, prairies, and deserts. According to Leopold, natural beauty is not a simple mirror of human notions of the sublime or picturesque. There is a spontaneous beauty in interconnected living things and their "fit" within a natural geography that goes beyond whether a landscape is worthy of a painting or a color slide. The beauty of a swamp or prairie involves the biological story it tells that includes both geology and humanity. In environmental philosopher Baird Callicott's apt phrase, "We cannot love cranes and hate marshes."[121] Ecology gives substance to scenery.

In a single summary statement Leopold wrote that "the practice of conservation must spring from a conviction of what is ethically and esthetically right, as well as what is economically expedient. A thing is right only when it tends to preserve the integrity, stability, and beauty of the community, and the community includes the soil, waters, fauna, and flora, as well as people."[122] Leopold urged us at least to keep all the parts while we are playing with nature. These missing parts might become important when we try to repair or reconstruct. As the philosopher Daniel Dennett puts it, this is not just a game, but a matter of life and death: "*Getting it right*, not making mistakes, has been of paramount importance to every living thing on this planet for more than three billion years."[123]

We learn how little we know about American history when we see it from an environmental framework. Every landscape, even when it is a palpable presence, still has its own code, dense with evidence, complex and cryptic.[124] Deborah Tall writes that when we bring an ecosystem into the foreground, we complete the story: "I need to learn the plot and poetry of this place, the outlines of time passing on it, in order that it not be merely scenery" or background noise.[125]

Reinhabiting Private Property

Private property remains the fundamental imprint upon America's geography. Private property defines American hubris toward the landscape. A person buys a piece of land, views it as an autonomous possession owned in its entirety, uses it profitably or wastefully, divides it up, sells it, or simply holds it indefinitely. For Americans the features and qualities of a piece

of land have disappeared underneath layers of human concep-
tions, such as marketable wealth. Law professor Eric Freyfogle
urges us to reorient our view, to begin with the land as its own
lawgiver: each tract of land contains its own internal limits and
opportunities. Its natural character cannot be improved on by
human efforts without equal or greater loss of its own identity, including durability. This builds upon Leopold's land ethic.
What is our circle of obligation, our territory of responsibility?
As Leopold learned, this means accepting certain limitations set
by the land and accepting full accountability for the land.[126]

Freyfogle employs ecological science to argue that nature's
communities are far more than just a collection of pieces that
can be acquired as autonomous tracts of land. These pieces
cannot randomly bounce around without this resulting in their
utter destruction. Instead "they fit together more snugly than
this, constituting an interdependent whole," complete unto itself prior to human intervention.[127] Freyfogle quotes a 1972
Wisconsin Supreme Court decision, *Just v. Narinette*: "An owner
of land has no absolute and unlimited right to change the essential natural character of his land so as to use it for a purpose
for which it was unsuited in its natural state and which injures
the rights of others."[128] The virtue of such restraint is that it
leaves the door open for the future well-being of both the land
and its human habitation.

Freyfogle asks for a revolutionary shift, that we look at nature
from the bottom up. It is destructive, he says, to have begun our
history by depending upon the notion of the efficient alloca-
tion of resources through marketplace economics. A tectonic
shift away from marketplace forces is the necessary next phase
of American history, "the resettlement of America," this time
in a knowledgeable and more durable manner, based on community rather than isolated parts.[129] Resisting American placelessness, Freyfogle urges permanence rather than transience.
Ownership through engagement with the ecological features

of one's place allows one's place to become the foundation for one's mind and heart. In my village, for example, our planning commission is working to secure such a foundation, in our case based in an oak forest rising from the Indiana Dunes lakeshore. Some members of my community instead see only autonomous private lots holding freestanding structures—plats and buildings—that are disconnected from the interdependent whole.

Ted Steinberg, environmental historian and law professor, has examined the implications of traditional property ownership. He finds that nature fights back: cycles of drought become Dust Bowls, polluted water never stays still, contaminated factory sites turn into Superfund black holes: "The natural world's continual resistance to human meddling suggests the weaknesses of a system of thought that centers so thoroughly on possession." It is foolhardy to impose "capitalistic logic on the seemingly nonideological matter-in-motion we call nature. The impulse to turn everything into property has not just confused but impoverished our relationship with the natural world."[130] He adds that we will not prosper by "reducing that world in all its complexity into a giant legal abstraction." Turning the land into property, which in turn becomes an abstraction, is the worst formulation of a virtual reality that will fail badly and carry us with it: "Property in land is an act of denial, a wish for closure, for solid control and dominion of nature."[131] Steinberg calls property a legal fiction that is hard to fathom.

Freyfogle admits that permanence instead of placelessness demands hard work, as well as a change of mind: "To stay home is to become attentive to a chosen place and let nature shape the methods and rhythms of everyday life . . . recognizing nature's constraints and peculiarities and living at peace with them." It demands knowing the specific ways of connecting to one's own place. One of these tools is through modifications in the law: "Property law, broadly conceived to include natural resources law, could improve greatly if it paid more attention to the land

and didn't focus so exclusively on the competing interests of people." Not that such a revolution would impoverish the property owner. Clear title to a piece of land, and the relative autonomy that comes with it, can bring positive outcomes. It can give the owner a sense of security over the long run. This in turn brings an incentive to care for the land, to restore it to ecological health—the ethic of stewardship. Freyfogle uses a fictional farmer, Mat, to make his point: "Would Mat behave differently if his land title were insecure? Surely he would invest less of himself in the place, be more inclined to push the land to its limits, exhaust the soil, log the woodlot, and ignore erosion."[132]

Leopold, Freyfogle, and Steinberg are not alone in their quest. The biologist Rene Dubos, in his pregnantly titled book *The Wooing of the Earth*, suggests that while every place on earth has been transformed by human presence, an ecological approach, looking beyond human self-interest, is more complex and far more appropriate to the natural world than traditional resource analysis. Intimacy with one's place, continues Dubos, offers a "quality of blessedness." This involves a sustained conversation between people and land: "When we impose our desires without regard for the qualities or needs of our place, then landscape may be cursed rather than blessed by our presence."[133] On the other hand, an ordinary place, properly inhabited, becomes a sacred place.

"Leopold's Children" have been equally impassioned. Writer and farmer Wendell Berry seeks "an ecological intelligence" that especially revokes rampant individualism: "a sense of the impossibility of acting or living alone or solely in one's own behalf."[134] Both Steinberg and Freyfogle agree with Thoreau: "Man is rich in proportion to the number of things he can afford to let alone." The United States, by any measure, enjoys a rich economy, but it is running headlong into a fatally weakened geography, as long as we persist in being sleepwalkers in paradise.[135]

We seek permanence that we cannot find in the flood of cyber-space. The total spectrum of human meaning—personal, his-torical, social, political, economic, cultural—becomes concen-trated in a single place. German philosopher Martin Heidegger, and his disciple Japanese philosopher Watsuji Tetsuro, looked to the security of an "existential foothold," a highly personal anchor for each of us, as a *Dasein*, a word that is literally trans-lated as "being there."[136] Echoing Mircea Eliade's hierophany, Mexican professor of aesthetics and design Katya Mandoki writes: "Power is believed to be a magical and superhuman force emanating from a site, rather than the effect of human will or physical might. . . . Taking hold of a place is taking hold of pow-er. Such a place, like a fetish, endows its occupants with an aura of worldly or divine power. . . . Yet the place, as a sponge that has absorbed layers of time and history, is ultimate."[137] Yi-Fu Tuan finds such a place filled with an independent presence.[138] Place has particularity. It is peculiar to itself. Place is both irre-ducible to anything less and impossible to contain in human terms. Particularity means new, original, underivable, separate, and spontaneous—*sui generis*. Particularity is the essence of both American space and American time. It reveals the faulty world we encounter when we act, for example, only as consumers, in-stead laying bare deeper levels of satisfying meaning.

The connection goes both ways. We do not exist in our best places like a rock or a mailbox. Rather, says the French philoso-pher Maurice Merleau-Ponty, we inhabit or haunt our places; we are literally there. Our "presence" persists in our absence. The German sentence is "Es spunkt hier," referring to a pow-erful presence that inhabits a place in ways greater than its so-called natural condition.[139] Yi-Fu Tuan notes that we feel strong-ly compelled to change "amorphous space" into "articulated geography." The mental picture is superimposed upon physi-cal place: "We have an insider's view of human facts."[140] We humans seem unable to look at our worldly experiences entirely on a pragmatic level that is restricted to direct perception. Tuan adds, "The human being, by his mere presence, imposes a schema on space."[141] This is the fate, and the glory, attached to being human.

Authentic place carries an inner necessity, the certitude of first-time, firsthand insight. It has the immediacy of direct, empirical sensation. In paleontologist Catherine Badgley's words, "It's a strange moment when a place encountered for the first time has so many familiar qualities and details—of scale, texture, color, the scent of the air—that it seems to be the discovery of a long-lost home."[142] Yi-Fu Tuan celebrates "the first glimpse of the desert through a mountain pass or the first plunge into forested wilderness [that] can call forth . . . inexplicably, a sense of recognition as of a pristine and primordial world one has always known."[143] The key is to break through the veil of ordinary place into the plentitude of nature as "wholly other." Saint Augustine, in his *Confessions*, writes: "What is that which gleams through me and smites my heart without wounding it? I am both a-shudder and a-glow. A shudder, in so far as I am unlike it, a-glow in so far as I am like it."[144] Edmund Burke, in his eighteenth-century description of the sublime, added the important corollary that the power that emanates from an autonomous natural world profoundly enlarges the wonder of being human. Western historian Dan Flores reminds us that "knowledge is insufficient in itself to enable humans to open to a place as home. Knowledge is too cerebral. Experiential, sensuous immersion—the way we've always done it, is the path home."[145] Landscape historian John Stilgoe urges his readers simply to accept that "outside lies magic." Catherine Badgley identifies places of formative experiences: "I was also part of the ancient, important play, not just a distant observer. I saw there also the 'interbeing' of rocks, water, and life." She adds, "There I felt power, mystery, and honor in my life being anchored to the history of the earth . . . [the] face to face encounter with deep time, carrying everything along . . . linking the phases of my life to lichen time, thimbleberry time, earth time."[146] She let Nature take over.

In 1864, during the Civil War, the young John Muir—later the savior of Yosemite and still later the father of the Sierra Club—took off to Canada to avoid military conscription. While a conversion experience can easily be overdramatized, Muir went through a personal change when, on a hike above the Great Lakes, he collapsed into tears at the sudden sight of a rare exquisite wild orchid (*Calypso borealis*) in forlorn country. He discovered a seamless intermingling between the human self and the cosmos: "Presently you lose consciousness of your own separate existence: you blend with the landscape, and become part and parcel of nature."[147] His intense state of consciousness corresponded with a highlighted moment in the immediate natural world. Never far from his biblical background, Muir folded God into Nature in a kind of pantheism. Philosopher Edward S. Casey writes, "The sense of place that counts here is not that of place as it contains and perdures but as it lights up with the sudden spark of a single striking image, like a shooting star in the dark abysm of night."[148] And Aldo Leopold, whose land ethic dominates modern environmentalism, found himself transfixed, and forever changed, when he looked into the dying gaze of a wolf and saw an unexpected nonhuman presence.[149]

The modern nature writer Annie Dillard has spent her entire writing career trying to push aside the veil. In her 1974 book *Pilgrim at Tinker Creek* she writes, for example, of a dazzlingly lit tree whose appearance echoes that of Moses's burning bush: "One day I was walking along Tinker Creek thinking of nothing at all and I saw the tree with the lights in it. I saw the backyard cedar where the mourning doves roost charged and transfigured, each cell buzzing with flame." It was as if she could see, fully displayed, the organic mechanics that gave the tree its life force: "It was less like seeing than being for the first time seen, knocked breathless by a powerful glance. The flood of fire abated, but I'm still spending the power."[150] Dillard adds: "I saw the cells in the cedar tree pulse, charged like wings beating praise. Now, it would be too facile to pull everything out of the hat and say that mystery vanquishes knowledge . . . that the frayed and nibbled fringe of the world is a tallith, a prayer shawl . . . the new is always present simultaneously with the old, however hidden. The tree with the lights in it does not go out; that the light still shines on an old world, now feebly, now bright." She struggles with words to describe the sensation free of human barriers: "I am a sacrifice bound with cords to the horns of the world's rock altar. . . . A sense of the real exults me; the cords loose; I walk on my way."[151]

In the same vein Merleau-Ponty wrote: "As I contemplate the blue of the sky I am not set over against it as a cosmic subject; I do not possess it in thought, or spread out towards it some idea of blue such as might reveal the secret of it. I abandon myself to it and plunge into the mystery, it 'thinks itself within me.' I am the sky itself as it is drawn together and unified and as it begins to exist for itself; my consciousness is saturated with this limitless blue."[152] The anchor of place grants our foothold in authenticity. Whatever else can be said about our individual humanity, physical location is the most direct human experience—our primary world, the extraordinary home base of sensations, with the immediacy of *here*. It is direct, primordial, and spontaneous. The philosopher Hans Jonas reminds us that we live continuously, minute-by-minute, hour-by-hour, day-by-day, because our metabolism continuously interacts with its environment to receive necessary supplies and information. The organic life process as much as anything else defines humans.[153]

We can wake from our sleepwalking. Place can be intimate, intense, and robustly real.[154] When we see ourselves as inhabitants instead of as drifters, we have an immediate and primary reference to an objective presence, an object outside the self, a primary datum. This is the authenticity that we are looking for.

When a place has an enduring quality, when it is the container of intense power, it can be described as a *sacred* place. We have already seen how Americans (and pilgrims around the world) seek out holy places for personal, religious, and patriotic reasons. Over time more and more American places have received sacred status. Regions like New England, or the South, or the Midwest, have become "centers" of an American identity: we speak piously of our "heartland." The West receives special attention because we see it as closer to the physical environment—mountain snows and desert desiccation and ocean edges carrying us beyond the horizon.[155] A sacred place is larger than life, containing more than what we construct in our own imaginations.

Certain places are conceived as having more magical power than others, whether sacred places like the Lincoln Memorial or demonic places like the impending nuclear-waste site at Yucca Flats in Nevada. As humans we can demonize places, like Buffalo's Love Canal, or sacramentalize them, like Yosemite. Political power may change, religious values collapse, the language be replaced, but the site remains immutable throughout time. Mircea Eliade and Yi-Fu Tuan emphasize that in our sacred places we can abolish time, resist facile change, and reach beyond physical limits. We seek to carve out "permanences"—sacred or sacramental places—from the chaotic and indifferent flow of nature.

Yi-Fu Tuan calls our preoccupation with the discovery of sacred places *topophilia*. Geographer John K. Wright calls it *geopiety*. The American philosopher-theologian William James found in sacred places "an everlasting presence," both intimate and alien, both terrible and lovable. The twentieth-century theologian Rudolph Otto wrote of the *mysterium tremendum* contained in unique holy places, emphasizing the profoundly disturbing encounter with the Other. Mircea Eliade describes passage between cosmic regions. We can pass beyond earthly place to another realm where the usual limits no longer exist: "Every sacred spaces implies a hierophany, an irruption of the sacred that results in detaching a territory from the surrounding cosmic milieu and making it qualitatively different."[156]

The psychologist Carl Jung even identified certain places that possess an "autonomous numinosity," such as the region around Taos, New Mexico, and northern Arizona's Sedona.[157] The popular theologian Joseph Campbell sums the notion up: "The idea of a sacred place where the walls and laws of the temporal world may dissolve to reveal a wonder is apparently as old as the human race." Echoing Eliade and Michael Cohen, Campbell continues, "It is as if the human psyche were continually feeling along the surface of a great rock face, in search of the slightest fissure, a discontinuity that might afford entry behind the rock to a numinal reality which both underlay and transcended the stone facade."[158] My cognitive-mapping students come to believe in Tuan's topophilia, Wright's geopiety, and perhaps even Otto's *mysterium tremendum* without having known of Tuan, Wright, or Otto.

Theologian Belden C. Lane asked, in this light, how we can "recover enchantment without syrupy nature mysticism." German sociologist Max Weber had earlier written of our Western disenchantment with the notion of any sacred aura in our places. This disenchantment helped establish the modern relationship to nature, which left it open to technological manipulation and a sense of triumph over nature, stripped of magic and mystery.[159] Economist Michael Pollan concludes that we have gone so far in uprooting nature that we can no longer retreat to some idealized communion with the natural world: "The doors to Eden have closed."[160] Lane instead finds an answer in theologian Paul Ricoeur's demand for a modern critical analysis. We can move away, says Ricoeur, from our original

innocent naiveté through intellectual objectivity and demythol-
ogizing, toward a far more sophisticated "second-naiveté," in
which wonder is restored.[161] Some of this critical analysis has
resulted in the interdisciplinary science of ecology and the land
ethic of Aldo Leopold. Others look to the eyes-wide-open encounters described with passion and rigor by Annie Dillard.

Let us admit that place can seem arbitrary, a momentary
pause in the constant flow of an undetermined world. This is
both the ideal and the flaw of American history. Daniel Boorstin
sees such vagueness as a crucial American virtue. Donald Meinig
reminds us, "Each of us creates and accumulates places out of
living whenever we pierce the infinite blur of the world and fix
a piece of our environment as something distinct and memorable."[162] Our physical place is not only filled with lumpiness
and artifacts but is the true repository, in Deborah Tall's words,
of "cultural and spiritual revelations . . . as vivid as the rocks
and trees you can touch, the past made visible . . . and the future into which you can confidently walk."[163] There is no croco-
dile pit outside that door. Indeed, we can speed through the
black hole—Engineered, Consuming, Triumphal America—to
the other side. Our permanent place is trustworthy. The total
spectrum of human meaning—personal, historical, sacred—im-
plodes continuously toward our single place.

Environmental science breaks through into this other world.
Because it simultaneously emphasizes a new knowledge, admits
its imperfect method, and acknowledges how little we know, it
teaches us that place, and our inhabitation of place, cannot be
exhausted. Aldo Leopold and Wendell Berry have long called
for an "ecological intelligence" that leads to actions that are
proper to one's place. Place is palpable but beyond final definition. It may well remain perpetually inexpressible, eluding total
apprehension. There are hidden forces in nature that seem in-
calculable, arbitrary, even mere caprice, unknowable to us and
indifferent to us. Nature truly is the Other, with qualities that

are definitive in themselves. Place is supramundane. Loss of place—placelessness—would indeed be loss of essence.

The challenge for Americans is to dwell intimately with nature, participating in the "quality of blessedness" that makes us revere it. In his essays on home and place Scott Russell Sanders calls on Americans to reverse their historical momentum—to become inhabitants rather than drifters, to shift from devouring to healing. The challenge is to render place meaningful even while our contemporary world rushes toward placelessness. By being inhabitants, we have the power to translate a location, changing it from a site measured merely in dollar signs into a place that is durable, reliable, authentic, and thus nurturing. Casey notes: "To dwell is to exercise patience-of-place; it requires willingness to cultivate, often seemingly endlessly, the inhabitational possibilities of a particular residence. Such willingness shows that we care about *how* we live in that residence that we care about it as a place for living well."[164] The German philosopher Edmund Husserl insisted that without each of us having a dynamic, lived place in which to settle, things would become "free-floating, flying off in all directions."[165] Like the Appalachian mother we must stomp our feet and refute the abstracted dot on the map.

Illustration Acknowledgments

Frontispiece A. David Johnson, *The Book as Landscape*. Cover illustration, New York Times *Book Review*, March 20, 1983.

Frontispiece B. Sidney Harris, *Spacious Skies . . . Leo's Diner*. © The New Yorker Collection 1981 Sidney Harris from cartoonbank.com. All Rights Reserved.

1. Frances F. Palmer, *The Farmers Home—Harvest* (1864). Currier & Ives. Lithograph with hand coloring. Fine Arts Museum of San Francisco. Gift of Joseph Martin Jr., 1994.120.37. Reprinted by permission.

2. Thomas Moran, *Mountain of the Holy Cross* (1875). Oil on canvas, 82⅛ x 64¼ in. Autry Museum of Western Heritage, Los Angeles. Reprinted by permission.

3. John Opie, *Snake River in the Tetons: Ansel's View* (2006). Photograph. Copyright 2006 by John Opie.

4. Young Hyun, *3-D Hyperbolic Visualization of Internet Topologies* (2000). Cooperative Association for Internet Data Analysis—CAIDA (http://www.caida.org/~youngh/walrus/walrus.html). Copyright 2003, The Regents of the University of California. Reprinted by permission.

5. Tim Bray, *Visualizing Web Space as a 3-D Cityscape* (November 2000). Illustration from Martin Dodge and Rob Kitchin, *Atlas of Cyberspace* (Harlow, England: Addison-Wesley, 2001), 146.

6. *Sims 2 Open for Business*. Electronic Arts (http://thesims2.ea.com/about/ep3_screenshots.php). Reprinted by permission.

7. Thomas Cole, *View from Mount Holyoke, Northampton, Massachusetts, after a Thunderstorm—The Oxbow* (1836). Oil on canvas, 51½ x 76 in. The Metropolitan Museum of Art. Gift of Mrs. Russell Sage, 1908 (08.228). © The Metropolitan Museum of Art. Reprinted by permission.

8. Erwin J. Raisz, detail from *American Grand Tour*, landform map of the United States (http://www.raiszmaps.com). Reprinted by permission.

9. Catskill Mountain railway station, Haines Corners, New York (1902?). Photograph. Library of Congress, Prints and Photographs Division, Detroit Publishing Company. Public domain.

10. L. C. McClure, *Busy Day at Soda Spring, Manitou, Colorado* (ca. 1900). Photograph. Western History/Genealogy Department, Denver Public Library. Reprinted by permission.

11. (Karl) William Hahn, *Yosemite Valley from Glacier Point* (1874). Oil on canvas, 34 x 53 in. California Historical Society. Gift of Albert M. Bender, acc# X57–548–1-2. Reprinted by permission.

12. G. E. Curtis, *Horseshoe Fall, Moonlight, No. 79* (1870). Stereoscopic image of photograph. Author's collection.

13. Albert Bierstadt, 1830–1902, *Among the Sierra Nevada, California* (1868). Oil on canvas, 72 x 120 in. Smithsonian American Art Museum. Bequest of Helen Huntington Hull, granddaughter of William Brown Dinsmore, who acquired the painting in 1873 for "The Locusts," the family estate in Duchess County, New York. Reprinted by permission.

14. Thomas Cole, *Falls of the Kaaterskill* (1826). Oil on canvas, 43 x 36 in. The Westervelt-Warner Collection of Gulf States Paper Corporation, Westervelt-Warner Museum of American Art, Tuscaloosa, Alabama. Reprinted by permission.

15. Engraving of Yosemite Valley (1874). Source unknown. Author's collection.

16. Yellowstone Inn and auto (1922). Photograph. Yellowstone National Park Photograph (3295). National Park Service, USDI. Public domain.

17a. John Opie, Shiprock, a volcanic plug near the Four Corners region (2006). Photograph. Copyright 2006 by John Opie.

17b. John Opie, desert scene in southeastern Utah (2006). Photograph. Copyright 2006 by John Opie.

18. William Henry Holmes, *Vishnu's Temple*. Illustration from J. W. Powell, *The Exploration of the Colorado River and Its Canyons* (1895; reprint, New York: Dover, 1961), 396. Public domain.

19. Administration Building and Main Basin at Night, World Columbian Exposition (1893). Chicago Historical Society, ICHi-02554. Reprinted by permission.

20. *The Corliss Engine, in Machinery Hall.* From *The Illustrated History of the Centennial Exhibition* (1876; reprint, Philadelphia: National Publishing Company, 1975), 158. Public domain.

21a. James Stevenson, cartoon referring to the 1893 World's Columbian Exposition. *New Yorker Magazine*, March 17, 1980. Reprinted by permission, cartoonbank.com.

21b. James Stevenson, cartoon of the entrance to Disneyworld. *New Yorker Magazine*, October 23, 1971. Reprinted by permission, cartoonbank.com.

22. *Sim City.* Electronic Arts (http://thesims2.ea.com/about/screenshots.php). Reprinted by permission.

23. Elizabeth Parsons and Rosemary Atwater, *New York and Brooklyn* (1875). Currier & Ives. Library of Congress, Geography and Map Division. Public domain.

24. H. D. Manning, *A Midnight Race on the Mississippi* (1860). Currier & Ives. Lithograph with hand coloring. 22.9cm x 33.4cm. F. F. Palmer, del. Fine Arts Museum of San Francisco. Gift of Anne Hoopes in memory of Edgar M. Hoopes III, 1989.1.56. Reprinted with permission.
F. F. Palmer, del., *The "Lightning Express" Trains: Leaving the Junction* (1863). Currier & Ives. Museum of the City of New York, 1220 Fifth Ave. at 103rd St., New York NY. Public domain.

25. Map showing the Toledo, Ann Arbor, and Grand Trunk Railway and its connections (1881). "Railroad Maps, 1828–1900," Library of Congress, American Memory Map Collection. Public domain.

26. Adolph Karst, after J. D. Woodward, *Superior Street, Cleveland, from*

Presbyterian Church (1873). Wood engraving. From *Picturesque America* (1872–74). Public domain.

The "Town Pump" (1908). From the Russell Sage Foundation, *The Pittsburgh District: Civic Frontage* (New York: Survey Associates, 1914), 131. Public domain.

27. George Caleb Bingham, *Daniel Boone Escorting Settlers through the Cumberland Gap* (1851–1852). Oil on Canvas, 36½" x 50¼". Mildred Lane Kemper Art Museum, Washington University in St. Louis. Gift of Nathaniel Phillips, 1890.

28. John Gast, *Westward Ho (American Progress)* (1872). Library of Congress, Prints and Photographs Division (LC-USZ62-737). Public domain.

29. Student cognitive map of Fair Oaks, California (ca. 2001). Author's collection.

30. Fanny F. Palmer, del., *The Rocky Mountains. Emigrants Crossing the Plains* (1868). Currier & Ives. Library of Congress, Prints and Photographs Division (LC-USCZ2-3758). Public domain.

31. Claude Fiddler, *Bristlecone Pines (partial)* (2001). Photograph. Used by permission.

32. Valerie Cohen, *Bristlecone Pines* (1998). Watercolor. Used by permission.

Maps

1. Map of U.S. Counties, National Atlas of the United States (2005). United States Geological Survey, USDI, Washington DC (http://nationalatlas.gov/natlas/Natlasstart.asp). Public domain.

2. Hydrological Unit Boundaries Map 2812 (April 1998). Natural Resource Conservation Service, U.S. Department of Agriculture. Public domain.

3. Major Land Resource Area (MLRA) Boundaries Map 2147 (July 1998). National Resource Conservation Service, U.S. Department of Agriculture. Public domain.

4. Robert G. Bailey, *Ecoregions and Metropolitan Areas* (2001). In *Ecoregion-Based Design for Sustainability* (New York: Springer-Verlag, 2001), 108. Reprinted by permission of author.

5. Greater Yellowstone Ecosystem, Greater Yellowstone Coalition (2005). Ecosystem Map © Greater Yellowstone Coalition (http://www.greateryellowstone.org). Used with permission.

Notes

Introduction

1. The definition is based on the on-line encyclopedia Wikipedia (http://www.wikipedia.org), which is continuously interactive and not always trustworthy. Its use, though, is appropriate in this context.
2. Hamilton, "Theoretical Physics."
3. Other examples come to mind, including Alice's fall into Wonderland and Dorothy's whirlwind journey from black-and-white Kansas to the Technicolor Land of Oz.
4. Hamilton, "Theoretical Physics."
5. See the discussion on Freud, Proust, and Saint Augustine in Sacks, *Anthropologist on Mars*, 172.
6. Tuan, *Space and Place*, 82–83.
7. Tuan, *Space and Place*, 69–73, 79, 82, 127.
8. Tuan, *Space and Place*, 82–83; see also 85–87.
9. Kimball, "Fortunes of Permanence." This on-line article is also the last chapter of Kimball's book *The Survival of Culture*; see p. 14.
10. Qtd. in Kane, "Man That Got Away."
11. Dillard, *Pilgrim at Tinker Creek*, esp. 124–28, 142–45, 271–74. See also Stilgoe, *Outside Lies Magic*; Eiseley, *Invisible Pyramid*, 40.

12. Tall, *From Where We Stand*, 8, 17.

13. Sanders, *Country of Language*, 35–36, 40.

14. Stilgoe, *Outside Lies Magic*, 8–9.

15. Stilgoe, *Outside Lies Magic*, 181, 185.

16. See Greene, "One Hundred Years of Uncertainty."

Chapter 1. Welcome to VirtuaLand

1. Dodge and Kitchin, *Atlas of Cyberspace*, 154.

2. Qtd. in Greenland, "Nod to the Apocalypse," 5.

3. Turkle, *Life in the Screen*, 236.

4. Lindgren, "Generation Xbox," 8.

5. This is the thesis of Gelernter, "Second Coming."

6. Turkle, *Life in the Screen*, 13.

7. Dodge and Kitchin, *Atlas of Cyberspace*, 3.

8. Leland, "Family's Choices."

9. Qtd. in Caro, "Living Room's the Battlefield."

10. Reed Johnson, "Mass Media's Last Blast."

11. Qtd. in Reed Johnson, "Mass Media's Last Blast."

12. Dodge and Kitchin, *Atlas of Cyberspace*, 2.

13. Lessig, "Internet under Siege," 3; see also "Seven Questions."

14. See Thompson, "Game Theories."

15. Lindgren, "Generation Xbox," 8.

16. See Wertheim, *Pearly Gates of Cyberspace*, 158–214.

17. Qtd. in "Will the Internet Change Humanity?" 3, 1.

18. Dodge and Kitchin, *Atlas of Cyberspace*, 2.

19. Qtd. in Thompson, "Game Theories."

20. See the discussion in Armstrong, *Short History of Myth*, 65, 69–71.

21. See Jones, "Storybook Ending"; Eisenberg, "World through PC-Powered Glasses."

22. Jones, "Storybook Ending."

23. Lindgren, "Generation Xbox," 8.

24. Qtd. in "Multitasking Generation," 53.

25. "Multitasking Generation," 53.

26. Schiesel, "Welcome to the New Dollhouse." See also Will Wright, "Dream Machines."

27. Steven Johnson, "Don't Fear the Digital," 56.

28. See Huizinga's classic study *Homo Ludens*, notably the definitions and cultural context provided in "The Nature and Significance of Play," 1–27.

29. Huizinga, *Homo Ludens*, 4.

30. Huizinga, *Homo Ludens*, 8.

31. "Worlds without End."

32. Turkle, *Life in the Screen*, 13, 14; see also 178, 267–68; Sacks, *Anthropologist on Mars*, 35.

33. See Castronova discussed in Thompson, "Game Theories," 1–4.

34. Qtd. in, Thompson, "Game Theories," 4.

35. Thompson, "Game Theories," 4.

36. Qtd. in, Thompson, "Game Theories," 22.

37. Gimblett et al., "Intelligent Agent Based Model"; see also Ball, "Mathematicians Find Path Less Traveled."

38. Turkle, *Life in the Screen*, 263.

39. See Schiesel, "Video Games Are Their Major."

40. Keller, "Plugged-in Proust."

41. Dodge and Kitchin, *Atlas of Cyberspace*, 251.

42. Qtd. in Leland, "Pixel Canvas."

43. See the discussion in Overbye, "Quantum Trickery."

44. See Rush, "In Love with Reality"; see also Dodge and Kitchin, *Atlas of Cyberspace*, 61.

45. See Heffernan, "Climb In."

46. Qtd. in Rush, "In Love with Reality."

47. Qtd. in Keller, "Plugged-in Proust."

48. A quick scan of these endnotes demonstrates this irony.

49. Merchant, *Reinventing Eden*, 11–12 (e.g.); cf. Armstrong, *Short History of Myth*, 14–15.

50. See, e.g., Fixmer, "Soul Amitai Etzioni"; Lyman, "Virtual Reality Comes Back in New Guise"; O'Connell, "Mining the Minds of the Masses."

51. Wertheim, *Pearly Gates of Cyberspace*, 223, 258, 76, 82, 51; see also 214.

52. See Armstrong, *Short History of Myth*, 32–35.

53. Qtd. in Lehmann, "Hit the Road to Dreamland."

54. Koppell, "No 'There' Here," 16–18.

55. Armstrong, *Short History of Myth*, 4.

56. Emerson, *Nature*, 8.

57. Albanese, *Corresponding Motion*, 5; see also 4–7, 28–30, 170–73.

58. Qtd. in Wilson, *Romantic Turbulence*, 109; cf. Armstrong, *Short History of Myth*, 95.

59. Wilson, *Romantic Turbulence*, 111.

60. Wilson, *Romantic Turbulence*, 132.

61. Qtd. in Wilson, *Romantic Turbulence*, 139.

62. Wilson, *Romantic Turbulence*, 118, 119.

63. "Will the Internet Change Humanity?"

64. See the discussion in Jakle, *Tourist*, 286–89.

65. Jakle, *Tourist*, 286–89.

66. Qtd. in Sacks, "Mind's Eye," 56, 57.

67. Sacks, "Mind's Eye," 57.

68. Sacks, *Anthropologist on Mars*, 35, 23.

69. Sacks, *Anthropologist on Mars*, 133, 114.

70. Qtd. in Jakle, *Tourist*, 41

71. Sacks, *Anthropologist on Mars*, 49, 55, 56.

72. Sacks, *Anthropologist on Mars*, 58–59.

73. Kimball, "Fortunes of Permanence," 11.

74. Sacks, *Anthropologist on Mars*, 105.

75. Jackson, "Jefferson, Thoreau and After."

76. Kimball, "Fortunes of Permanence," 12.

77. Mazur, *Hazardous Inquiry*, 5.

78. Fortunato, review, 1–2.

79. Sacks, *Anthropologist on Mars*, 254.

80. Sacks, *Anthropologist on Mars*, 252.

81. Armstrong, *Short History of Myth*, 16–23.

82. Conron, *American Landscape*, xxiv.

83. Qtd. in Jakle, *Tourist*, 41–42.

84. Qtd. in Jakle, *Tourist*, 65–66.

85. Gopnik, "Unreal Thing," 69.

86. Gopnik, "Unreal Thing," 73.

87. Turkle, *Life in the Screen*, 10, 14, 246.

88. "Will the Internet Change Humanity?"

89. Sacks, *Anthropologist on Mars*, 164–165n.

90. Qtd. in Sacks, *Anthropologist on Mars*, 164n.

91. Kimball, "Fortunes of Permanence," 16.

92. Kimball, "Fortunes of Permanence," 11.

93. See the discussion of "modernity" in Armstrong, *Short History of Myth*, 137–38, 142–46.

94. Sacks, *Anthropologist on Mars*, 173–77.

95. Qtd. in Turkle, *Life in the Screen*, 22.

96. Sacks, *Anthropologist on Mars*, 35–36, 34n.

97. "Will the Internet Change Humanity?"

98. Kimball, "Fortunes of Permanence," 4, 21.

99. Turkle, *Life in the Screen*, 26; see also 47.

100. Hannah Arendt, *Crisis in Culture*, qtd. in Kimball, "Fortunes of Permanence," 4, 6.

Chapter 2. Antique America

1. See, e.g., Opie, "Seeing Desert as Wilderness."

2. Schaub, "Eyes Wide Open," 10

3. Eco, *Travels in Hyperreality*, 7–8, 16–18.

4. Fitzgerald, *Great Gatsby*, 101.

5. Lankford, *Captain John Smith's America*, 130.

6. Dickens, *American Notes*, 91.

7. Qtd. in McShine, *Natural Paradise*, 69.

8. Eliade, *Myth and Reality*, 139.

9. Leopold, "Wilderness," in part 3, "The Upshot," *Sand County Almanac*, 196.

10. Qtd. in McShine, *Natural Paradise*, 71.

11. Qtd. in Aron, *Working at Play*, 141, 131.

12. See the discussion in Ritzer and Liska, "'McDisneyization' and 'Post-Tourism.'"

13. See Grochowski, "Vegas' Virtual Reality."

14. Phil Tippett interviewed in Boal, "In the Tracks of *Jurassic Park*," 258.

15. *New York Times*, Oct. 7, 2001.

16. Aron, *Working at Play*, 132.

17. See the discussion in Aron, *Working at Play*, 286–89.

18. See the statistics and discussion in Aron, *Working at Play*, 146–47.

19. Aron, *Working at Play*, 152, 151.

20. See the discussion in Aron, *Working at Play*, 237–57.

21. Qtd. in Aron, *Working at Play*, 255.

22. Qtd. in Aron, *Working at Play*, 242.

23. Qtd. in Brown, *Inventing New England*, 144–45.

24. Brown, *Inventing New England*, 155–62.

25. Qtd. in Brown, *Inventing New England*, 141.

26. See Brown, *Inventing New England*, 75–76, 105–34.

27. Brown, *Inventing New England*, 8–9.

28. Garreau, *Nine Nations of North America*.

29. Qtd. in Angela Miller, *Empire of the Eye*, 17. For a discussion of Cole's *Oxbow* see 39–49.

30. The following sections on landscape art and tourism are based on the author's numerous presentations since 1979 of a slide-based comparison between travel expectations and travel realities.

31. Qtd. in Novak, *Nature and Culture*, 38–39. Artists as spokespersons for a national identity are a major theme of Novak, *Nature and Culture*, esp. 21–75.

32. Angela Miller, *Empire of the Eye*, 20.

33. Thoreau, *Maine Woods*, 71; Muir, *Our National Parks*, 1; Morris, "To the Woods," 385.

34. Brown, *Inventing New England*, 51–52, 61–62, 225 n.24.

35. Angela Miller, *Empire of the Eye*, 18.

36. Boorstin, *Image*, 79–80.

37. Angela Miller, *Empire of the Eye*, 79–80.

38. Hyde, *American Vision*, 43.

39. Hyde, *American Vision*, 45.

40. Hyde, *American Vision*, 47.

41. Hyde, *American Vision*, 96.

42. Qtd. in Hyde, *American Vision*, 110.

43. Hyde, *American Vision*, 111.

44. Hyde, *American Vision*, 85.

45. Qtd. in Hyde, *American Vision*, 85.

46. See Hyde, *American Vision*, 17–18. Also see McShine, *Natural Paradise*, 74–80, 107–30; Brown, *Inventing New England*, 53–54; Angela Miller, *Empire of the Eye*, 249–53; Novak, *Nature and Culture*, 34–44.

47. Qtd. in Bonami, *Universal Experience*, 173.

48. Jefferson, "Jefferson to Maria Cosway," 627–28.

49. King, *Mountaineering in the Sierra Nevada*, 99; also qtd. in Novak, *Nature and Culture*, 151.

50. Hyde, *American Vision*, 162–74, 183–90.

51. Qtd. in Jakle, *Tourist*, 41.

52. Qtd. in Jakle, *Tourist*, 23.

53. Qtd. in Jakle, *Tourist*, 75–76.

54. Qtd. in Jakle, *Tourist*, 74.

55. Tuan, *Space and Place*, 122.

56. See the definitive essay: Didion, "John Wayne: A Love Song."

57. See Hyde, *American Vision*, 6.

58. Interview with the author, Canyonlands National Park, July 1975.

59. O'Keeffe, *Georgia O'Keeffe*, 12; also see a discussion in Opie, "Seeing Desert as Wilderness."

60. Qtd. in McShine, *Natural Paradise*, 115.

61. Qtd. in McShine, *Natural Paradise*, 119.

62. Qtd. in Pomeroy, *In Search of the Golden West*, 159.

63. Pyne, *How the Canyon Became Grand*, 89.

64. Pyne, *How the Canyon Became Grand*, 91.

65. Hyde, *American Vision*, 191, 211; see also Goetzmann, *Exploration and Empire*.

66. Qtd. in Jakle, *Tourist*, 49.

67. Pomeroy, *In Search of the Golden West*, 94–104.

68. Qtd. in Aron, *Working at Play*, 230–31.

69. Lemons, in the draft article "National Park Management and Values," devises an extensive list of thirteen values or activities associated with the National Park System. Of these only four—life support, scientific, genetic diversity, intrinsic—could be considered ecosystem values. The others are market, recreational, aesthetic, cultural symbolization, historical, character-building, therapeutic, sacramental, and aspirational.

70. Muir, "Wild Parks and Forest Reservations"; see also Runte, *National Parks*, 65–67.

71. Growing public awareness of the nature of park ecosystems, and human impacts, is reflected in Stevens, "Latest Threat to Yellowstone."

Chapter 3. Human Kodaks in the Future Perfect

1. Much of the documentation for this chapter is drawn from Rydell, *All the World's a Fair*, see esp. 237, 235.

2. Rydell, *All the World's a Fair*, 2, 4–5.

3. See the discussion in Opie, "Energy and the Rise of American Industrial Society," 1.

4. Opie, "Energy and the Rise of American Industrial Society," 4

5. Qtd. in Rydell, *All the World's a Fair*, 35.

6. Rydell, *All the World's a Fair*, 2, 4–5.

7. Adams, "Dynamo and the Virgin."

8. Qtd. in Rydell, *All the World's a Fair*, 11–14.

9. Qtd. in Rydell, *All the World's a Fair*, 39–41.

10. Rydell, *All the World's a Fair*, 35–36.

11. Rydell, *All the World's a Fair*, 11–14.

12. Qtd. in Rydell, *All the World's a Fair*, 39–41, 71.

13. Qtd. in Rydell, *All the World's a Fair*, 4.

14. See Kidd, *1930s* (accessed Apr. 26, 2002).

15. Qtd. in Rydell, *All the World's a Fair*, 3–4.

16. Qtd. in Rydell, *All the World's a Fair*, 47.

17. Qtd. in Rydell, *All the World's a Fair*, 186.

18. Qtd. in Rydell, *All the World's a Fair*, 105–7.

19. Qtd. in Rydell, *All the World's a Fair*, 154–59.

20. Qtd. in Rydell, *All the World's a Fair*, 220, 231.

21. See Donald L. Miller, *City of the Century*, 492–93.

22. Qtd. in Rydell, *All the World's a Fair*, 220.

23. Qtd. in Donald L. Miller, *City of the Century*, 494; see also Rydell, *All the World's a Fair*, 39.

24. Donald L. Miller, *City of the Century*, 488, 494.

25. See the discussion in Donald L. Miller, *City of the Century*, 496–99.

26. Qtd. in Rydell, *All the World's a Fair*, 43.

27. Herrick qtd. in Rydell, *All the World's a Fair*, 38; Satt qtd. in Donald L. Miller, *City of the Century*, 495.

28. Friedman, "Semiotics of SimCity," 2.

29. See the 1992 Maxis Software "Toys Catalog," 4.

30. Qtd. in Radosh, "Cyber City," 36, 39.

31. Radosh, "Cyber City," 39.

32. Qtd. in Weinberg, "Five and a Half Utopias," 114.

33. Rydell, *All the World's a Fair*, 46.

34. See http://xroads.virginia.edu/~1930s/display/39wf/history.htm (accessed Apr. 26, 2002).

35. Qtd. in Wolf, "Hidden Kingdom."

36. Qtd. in Thomas, *Walt Disney*, 338.

37. See the Web site "Waltopia: The Florida Project": http://www .waltopia.com/ (accessed Apr. 26, 2002).

38. See http://xroads.virginia.edu/~1930s/display/39wf/history.htm.

39. See http://xroads.virginia.edu/~1930s/display/39wf/history.htm.

40. Qtd. in Wolf, "Hidden Kingdom."

41. See Wolf, "Hidden Kingdom" (my italics).

42. Qtd. in Wolf, "Hidden Kingdom."

43. Wolf, "Hidden Kingdom."

44. See http://www.mallofamerica.com/about_moa_history.aspx; see also http://en.wikipedia.org/wiki/Mall_of_America (both accessed Aug. 25, 2006).

45. Qtd. in Weisberger, *New Industrial Society*, 11.

46. Qtd. in Opie, *Nature's Nation*, 236.

47. See Thomas Wright, "What's for Afters?"

Chapter 4. Sleepwalking in America

1. McCullough, *Great Bridge*, 24.

2. Qtd. in McCullough, *Great Bridge*, 24.

3. Qtd. in Hays, *Conservation and the Gospel of Efficiency*, 127.

4. McCullough, *Great Bridge*, 37.

5. Qtd. in McCullough, *Great Bridge*, 41.

6. See McCullough, *Great Bridge*, 90–102.

7. Qtd. in McCullough, *Great Bridge*, 93.

8. Boorstin, *Image*, 223–24, 241.

9. White, *Dynamo and Virgin Reconsidered*, 149.

10. See the discussion in Petroski, *Engineers of Dreams*.

11. Potter, *People of Plenty*, 85.

12. See the visual decoding of the built environment—"everything that isn't nature"—in Hayes, *Infrastructure*.

13. Garreau, "Call of Beauty."

14. Hays, *Conservation and the Gospel of Efficiency*, 91.

15. Qtd. in Goetzmann, *Exploration and Empire*, 271.

16. Qtd. in Winter, *Transportation Frontier*, 105.

17. Qtd. in Winter, *Transportation Frontier*, 107.

18. See Lewis, *Divided Highways*; Heppenheimer, "Rise of the Interstates."

19. Warner, *Urban Wilderness*, 271.

20. Abbey, *Abbey's Road*, 191.

21. Emerson, "Ode, Inscribed to William H. Channing," n.p.

22. Gordon, "We Are All Spendthrifts Now."

23. Green, *Uncertainty of Everyday Life*, 14.

24. Qtd. in Crocker, "Consumption and Well-Being," 13.

25. Rousseau, *Social Contract*, 216.

26. Horseman, *Frontier in the Formative Years*, 118, 115–19; Larkin, *Reshaping of Everyday Life*, 171–80.

27. Tocqueville, *Democracy in America*, 536.

28. Qtd. in Sagoff, *Ethics of Consumption*, 2.

29. Melville, *Moby Dick*, 183.

30. Parkes, *American Experience*, 269–70.

31. Parkes, *American Experience*, 270.

32. Qtd. in Parkes, *American Experience*, 279.

33. Persons, *American Minds*, 298–315.

34. Heilbroner and Singer, *Economic Transformation of America*, 6. This is a major theme of this seminal study.

35. Bazelon, "New Factor in American Society," 267.

36. Qtd. in Sagoff, *Ethics of Consumption*, 3.

37. Schor, "What's Wrong with Consumer Society," 37–38.

38. Cf. the discussion in Greider, "One World of Consumers," 34–36.

39. Terry, "Supper's Experiment"; "Wolcott, "In Search of the Ripe Stuff."

40. Terry, "Supper's Experiment."

41. Wolcott, "In Search of the Ripe Stuff."

42. Wolcott, "In Search of the Ripe Stuff."

43. Ecological footprints are statistical analyses intended to identify consumption of natural resources on either an individual or a national basis. They are seen as an alternative to Gross National Con-

sumption and other indicators. There are many Web sites on ecological footprints, including http://myfootprint.org/, sponsored by an Earth Day organization, and http://en.wikipedia.org/wiki/Ecological_footprint, which provides a good overview of related issues, Web sites, and books, including criticisms of the concept.

44. Merk, *Manifest Destiny and Mission in American History*, 51.
45. Qtd. in Tuveson, *Redeemer Nation*, frontispiece.
46. Merk, *Manifest Destiny and Mission in American History*, 27.
47. Qtd. in Merk, *Manifest Destiny and Mission in American History*, 32.
48. Stephanson, *Manifest Destiny*, 110.
49. Qtd. in Merk, *Manifest Destiny and Mission in American History*, 55.
50. Qtd. in Merk, *Manifest Destiny and Mission in American History*, 235–36n.
51. Qtd. in Merk, *Manifest Destiny and Mission in American History*, 200.
52. Merk, *Manifest Destiny and Mission in American History*, 3.
53. Qtd. in Merk, *Manifest Destiny and Mission in American History*, 21.
54. Qtd. in Merk, *Manifest Destiny and Mission in American History*, 50.
55. Qtd. in Merk, *Manifest Destiny and Mission in American History*, 28.
56. Merk, *Manifest Destiny and Mission in American History*, 41.
57. Brand, *Whole Earth Catalog*, 3.
58. See the discussion in Elshtain, "Two Cheers for Democracy."
59. *New York Morning News*, Oct. 13, 1845, qtd. in Merk, *Manifest Destiny and Mission in American History*, 25.
60. See the nineteenth-century American interpretation of the biblical book of Revelation in Tuveson, *Redeemer Nation*, 8–9; see also 12–13 for the impact of early Christian millennialism and 16–19 for Reformation millennialism.
61. Qtd. in Tuveson, *Redeemer Nation*, 197–98.
62. Qtd. in Tuveson, *Redeemer Nation*, 59.
63. Tuveson, *Redeemer Nation*, 59.
64. Qtd. in Tuveson, *Redeemer Nation*, 83.
65. Qtd. in Tuveson, *Redeemer Nation*, 156–57.
66. Qtd. in Tuveson, *Redeemer Nation*, 167, 138.
67. Qtd. in Tuveson, *Redeemer Nation*, vii; qtd. in Chace, "Tomorrow the World," 3.
68. This is a major thesis in Zimmerman, *First Great Triumph*. See also Chace, "Tomorrow the World."

69. Chace, "Tomorrow the World," 4.

70. Stephanson, *Manifest Destiny*, 121–24.

71. Lundberg, "World Revolution, American Plan," 38–46; also qtd. in Potter, *People of Plenty*, 136.

72. See, e.g., Wines, "Russia Finds Virtue."

73. "Defining Century in America."

74. Lavadie, "USA Not as Good as We Hoped."

75. "Defining Century in America."

76. Qtd. in Hodgson, "New Statesman Profile."

77. Hodgson, "New Statesman Profile."

78. Ash, "Peril of Too Much Power."

79. Harries, "Understanding America."

80. Gowan, "Calculus of Power," 2, 13, 14.

81. Lukacs, "It's the End of the Modern Age."

82. Lohr, "Welcome to the Internet."

83. Qtd. in Lohr, "Welcome to the Internet."

84. Qtd. in Lohr, "Welcome to the Internet."

Chapter 5. Finding Authenticity

1. Qtd. in Gazaway, *Longest Mile*, 15.

2. See the discussion in Casey, *Fate of Place*, 203–5, 221.

3. See the introduction to Meinig, *Interpretation of Ordinary Landscapes*, 3.

4. Pascal, *Pensées*, 95; Casey, *Fate of Place*, 194, 340–41.

5. Qtd. in Casey, *Fate of Place*, 299.

6. See the discussion in Shepard, "Virtually Hunting Reality."

7. Kirk Johnson, qtd. in Hamilton, "Theoretical Physics."

8. Meinig, "Beholding Eye," 46.

9. See the analysis of Dewey in Casey, *Getting Back into Place*, 30–31.

10. Casey, *Fate of Place*, 219.

11. Agee, *Death in the Family*, iii.

12. This is a major theme in Casey, *Fate of Place*: see 234. See also Casey's discussion of Kant, Husserl, Whitehead, Bachelard, and Merleau-Ponty regarding the concept of specific implacement, in contrast with today's physics of infinite space and serial time, propounded by Aristotle, Newton, and even—with important modifications—Einstein (*Fate of Place*, 332–34).

13. See, e.g., the discussion in Mandoki, "Sites of Symbolic Density," 86.

14. Mitchell, *Ceremonial Time*.

15. Casey, *Fate of Place*, 225 (italics in original).

16. Casey, *Getting Back into Place*, xvii.

17. Casey, *Fate of Place*, 241, 294; see also 277, 279.

18. Leopold, "Wilderness," 196.

19. See the discussion in Casey, *Fate of Place*, 336; see also 339.

20. See the discussion in Casey, *Fate of Place*, 5–7.

21. Perry Miller, *Nature's Nation* and *Errand into the Wilderness*. Both volumes are collections of independent essays, linked by common themes.

22. Tuan, *Space and Place*, 110.

23. The best evocation of World War II soldiers' need to sense "home" remains the classic writings of correspondent Ernie Pyle—e.g., in *Here Is Your War*.

24. Tuan, *Space and Place*, 150.

25. Tuan, *Space and Place*, 115–16.

26. Casey, *Fate of Place*, xiii–xiv.

27. Dodge and Kitchin, *Atlas of Cyberspace*, 154.

28. Tuan, *Space and Place*, 56–57.

29. Sobel, "Sky Above," 16.

30. Tuan, *Space and Place*, 9.

31. Dodge and Kitchin, *Atlas of Cyberspace*, 38.

32. See the extensive discussion in Percy, *Message in the Bottle*, 30 and throughout.

33. Casey, *Fate of Place*, 237; see also Casey, *Getting Back into Place*, xv, 37.

34. From *Four Quartets*, qtd. in Sanders, *Secrets of the Universe*, 86, in the context of a fresh look at home place.

35. See the discussion in Turner, *Spirit of Place*, 37.

36. Qtd. in Turner, *Spirit of Place*, 331.

37. Qtd. in Turner, *Spirit of Place*, 132.

38. Qtd. in Turner, *Spirit of Place*, 212.

39. Turner, *Spirit of Place*, 213.

40. Qtd. in Turner, *Spirit of Place*, 208, 213.

41. Qtd. in Turner, *Spirit of Place*, 67.

42. Turner, *Spirit of Place*, 66.

43. See Turner, *Spirit of Place*, 78.

44. Qtd. in Turner, *Spirit of Place*, 83.

45. Qtd. in Turner, *Spirit of Place*, 77.

46. Turner, *Spirit of Place*, 90.

47. Feder, *Life of Charles Ives*, 10.

48. Feder, *Life of Charles Ives*, 7–8.

49. From the album notes to *Songs*, lyrics in foldout.

50. See the program notes in *Orchestral Music of Charles Lives*.

51. See the program notes in *Orchestral Music of Charles Lives*.

52. Qtd. in Feder, *Life of Charles Ives*, 17.

53. See the program notes in *Three Places in New England*, 3–7.

54. See the analysis in Feder, *Life of Charles Ives*, 57–58.

55. Qtd. in Feder, *Life of Charles Ives*, 59.

56. Agee, *Let Us Now Praise Famous Men*.

57. McHarg, *Design with Nature*, 16.

58. See Mandoki, "Sites of Symbolic Density," 88–90.

59. See a similar discussion in Tuan, *Space and Place*, 149.

60. Norton and Hannon, "Democracy and Sense of Place Values."

61. Qtd. in Tall, *From Where We Stand*, 36.

62. Smith, Light, and Roberts, "Introduction," 2.

63. Qtd. in Tuan, "Desert and I."

64. Sanders, *Staying Put*, 104–5.

65. Boorstin, *Americans: The Democratic Experience*, 290.

66. See the discussion in Flores, *Natural West*, 91–92.

67. See the discussion in Sopher, "Landscape of Home," 135.

68. Qtd. in Sanders, *Staying Put*, 103–4.

69. Tall, *From Where We Stand*, 7–8, 194.

70. See Smith, Light, and Roberts, "Introduction," 5.

71. See the discussion of alienation from any sense of place in Percy, *Message in the Bottle*, esp. 20–26.

72. Tall, *From Where We Stand*, 8.

73. Tuan, *Space and Place*, 4.

74. Matthiessen, *Snow Leopard*, 40.

75. Tuan, *Space and Place*, 20.

76. Qtd. in Casey, *Getting Back into Place*, 38.

77. Heisenberg, *Physics and Philosophy*, 58; cf. Sanders, *Staying Put*, 129–34.

78. Shepard, "Virtually Hunting Reality," 25–26.

79. Lennon and McCartney, *Beatles: 1962–66*, disk 2, artwork G.

80. Qtd. in Rothstein, "Ingmar Bergman, Master Filmmaker, Dies at 89."

81. Jackson, "Order of a Landscape," 231–32.

82. Qtd. in Maplas, "Finding Place," 22–23.

83. Kundera, *The Book of Laughter and Forgetting*, 157.

84. Schulman, "Bristlecone Pine."

85. Cohen, *Garden of Bristlecones*.

86. Cohen, *Garden of Bristlecones*, 114–15.

87. Cohen, *Garden of Bristlecones*, 14.

88. Brodsky, *On Grief and Reason*, 225–26.

89. Cohen, *Garden of Bristlecones*, xix.

90. Qtd. in Cohen, *Garden of Bristlecones*, 194.

91. Cohen, *Garden of Bristlecones*, 186.

92. Cohen, *Garden of Bristlecones*, 8–10.

93. Cohen, *Garden of Bristlecones*, 186–207.

94. Qtd. in Cohen, *Garden of Bristlecones*, 210.

95. Qtd. in Cohen, *Garden of Bristlecones*, 211.

96. Cohen, *Garden of Bristlecones*, 214, 215, 237.

97. See Cohen, *Garden of Bristlecones*, 1.

98. Qtd. in Cohen, *Garden of Bristlecones*, 183.

99. Cohen, *Garden of Bristlecones*, 186–87.

100. Cohen, *Garden of Bristlecones*, 121, 10.

102. Cohen, *Garden of Bristlecones*, 193.

103. Cohen, *Garden of Bristlecones*, 14–15.

104. Cohen, *Garden of Bristlecones*, 15, 147.

105. Qtd. in Cohen, *Garden of Bristlecones*, 188.

106. Qtd. in Cohen, *Garden of Bristlecones*, 188.

107. Qtd. in Cohen, *Garden of Bristlecones*, 195.

108. Thanks to geographer Chris Mayda of Eastern Michigan University for this idea of fractal nature. See her "Geographing.com."

109. "Major Land Resource Area (MLRA) Boundaries."

110. Bailey, "Ecoregions Map of North America."

111. See the discussion in Flores, *Natural West*, 91–92.

112. Kemmis, "Learning to Think like a Region."

113. Flores, *Natural West*, 92, 95.

114. Flores, *Natural West*, 95.

115. Sanders, *Staying Put*, 163–64.

116. Turner, *Spirit of Place*, 40.

117. Turner, *Spirit of Place*, 157.

118. Turner, *Spirit of Place*, 295.

119. Freyfogle, *Bounded People*, 131.

120. Leopold, "The Land Ethic," in part 3, "The Upshot," *Sand County Almanac*, 216.

121. Callicott, "Land Aesthetic," 162.

122. Leopold, "Land Ethic," 224–25.

123. Dennett, "Postmodernism and Truth," 4.

124. See Peirce Lewis, "Axioms for Reading the Landscape," 6.

125. Tall, *From Where We Stand*, 17.

126. Tall, *From Where We Stand*, 47, 96.

127. Freyfogle, *Bounded People*, 56.

128. Qtd. in Freyfogel, *Bounded People*, 107; see also 42, 108.

129. Freyfogle, *Bounded People*, 109–13.

130. Steinberg, *Slide Mountain*, 10.

131. Steinberg, *Slide Mountain*, 10, 49–50.

132. Freyfogle, *Bounded People*, 116–18, 121, 136.

133. Qtd. in Sanders, *Staying Put*, 116.

134. Qtd. in Sanders, *Staying Put*, 116.

135. Steinberg, *Slide Mountain*, 176; Freyfogle, *Bounded People*, 176.

136. See the analysis in Lane, *Landscapes of the Sacred*, 7–10.

137. Mandoki, "Sites of Symbolic Density," 86; see also Tuan, *Space and Place*, 6.

138. Tuan, *Space and Place*, 114.

139. See the discussion of the "numinous" in Otto, *Idea of the Holy*, esp. 127.

140. Tuan, *Space and Place*, 82–83, 85–88.

141. Tuan, *Space and Place*, 36.

142. Badgley, "Your Real Destination," 20.

143. Tuan, *Space and Place*, 184.

144. Augustine of Hippo, *Confessions*, ii.9.1.

145. Flores, *Horizontal Yellow*, 176.

146. Badgley, "Your Real Destination," 28–29.

147. Qtd. in Wolfe, *Son of the Wilderness*, 93.

148. Casey, *Fate of Place*, 288.

149. Leopold, *Sand County Almanac*, 130.

150. Dillard, *Pilgrim at Tinker Creek*, 33.

151. Dillard, *Pilgrim at Tinker Creek*, 241–42.

152. Qtd. in Lane, *Landscapes of the Sacred*, 54.

153. Jonas, *Phenomenon of Life*.

154. See Casey, *Fate of Place*, 217.

155. Tuan, *Space and Place*, 99.

156. Qtd. in Lane, *Landscapes of the Sacred*, 46.

157. Lane, *Landscapes of the Sacred*, 46.

158. Qtd. in Lane, *Landscapes of the Sacred*, 20.

159. See Lane, *Landscapes of the Sacred*, 22–23.

160. Qtd. in Lane, *Landscapes of the Sacred*, 240.

161. See the discussion in Lane, *Landscapes of the Sacred*, 23–24.

162. Meinig, *Interpretation of Ordinary Landscapes*, 4.

163. Tall, *From Where We Stand*, 19.

164. Casey, *Getting Back into Place*, 174.

165. See Casey, *Fate of Place*, 226, 233; see also Casey, *Getting Back into Place*, 22.

Bibliography

Abbey, Edward. *Abbey's Road.* New York: Penguin Books, 1979.

Adams, Henry. "The Dynamo and the Virgin (1900)." In *The Education of Henry Adams: An Autobiography*, 379–90. Boston: Houghton Mifflin, 1918.

Agee, James. *A Death in the Family.* New York: McDowell, Obolensky, 1957.

———. *Let Us Now Praise Famous Men.* Boston: Houghton Mifflin Company, 1941.

Albanese, Catherine A. *Corresponding Motion: Transcendental Religion and the New America.* Philadelphia: Temple University Press, 1977.

Armstrong, Karen. *A Short History of Myth.* New York: Cannongate, 2005.

Aron, Cindy S. *Working at Play: A History of Vacations in the United States.* New York: Oxford University Press, 1999.

Ash, Timothy Garton. "The Peril of Too Much Power." *New York Times*, April 9, 2002.

Augustine. *Confessions.* New York: Oxford World's Classics, 1998.

Badgley, Catherine. "Your Real Destination." In Grese and Knott, *Reimagining Place*, 20–29.

Bailey, Robert G. "Ecoregions Map of North America." Miscellaneous Publication No. 1548. Washington DC: USDA, Forest Service, rev. 1997.

Ball, Philip. "Mathematicians Find Path Less Traveled: Simulation of Grand Canyon Rafting Helps Manage Overcrowding in the Wild." *Nature*, January 14, 2003.

Bazelon, David T. "The New Factor in American Society." In William R. Ewald Jr., ed., *Environment and Change: The Next Fifty Years*, 264–86. Bloomington: Indiana University Press, 1968.

Boal, Ian A. "In the Tracks of *Jurassic Park*." In James Brook and Iain A. Boal, eds., *Resisting the Virtual Life: The Culture and Politics of Information*, 258. San Francisco: City Lights, 1995.

Bonami, Francesco, curator. *Universal Experience: Art, Life, and the Tourist's Eye*. Exhibition catalog. Chicago: Museum of Contemporary Art, 2005.

Boorstin, Daniel J. *The Americans: The Democratic Experience*. New York: Random House, 1973.

———. *The Image: A Guide to Pseudo-Events in America*. New York: Harper Colophon Books, 1964.

Brand, Stewart. *Whole Earth Catalog*. Menlo Park: Portola Institute, 1968.

Brodsky, Joseph. *On Grief and Reason*. New York: Farrar, Straus, Giroux, 1995.

Brown, Dona. *Inventing New England: Regional Tourism in the Nineteenth Century*. Washington DC: Smithsonian Institution Press, 1995.

Callicott, J. Baird. "The Land Aesthetic." In J. Baird Callicott, ed., *Companion to* A Sand Country Almanac, 157–71. Madison: University of Wisconsin Press, 1987.

Caro, Mark. "Living Room's the Battlefield in Digital Tech Wars." *Chicago Tribune*, November 27, 2005.

Casey, Edward S. *The Fate of Place: A Philosophical History*. Berkeley: University of California Press, 1997.

———. *Getting Back into Place: Toward a Renewed Understanding of the Place-World*. Bloomington: Indiana University Press, 1993.

Chace, James. "Tomorrow the World." *New York Review of Books*, November 21, 2002. http://www.nybooks.com/articles/15818.

Cohen, Michael P. *A Garden of Bristlecones: Tales of Change in the Great Basin*. Reno: University of Nevada Press, 1998.

Conron, John, ed. *The American Landscape: A Critical Anthology of Prose and Poetry.* New York: Oxford University Press, 1974.

Crocker, David A. "Consumption and Well-Being." In Mark Sagoff, *The Ethics of Consumption.* College Park MD: Institute for Philosophy and Public Policy, 1995.

Daly, Herman E., and John B. Cobb Jr. *For the Common Good: Redirecting the Economy toward Community, the Environment, and a Sustainable Future.* Boston: Beacon Press, 1994.

"A Defining Century in America." *New York Times,* December 13, 1999.

Dennett, Daniel. "Postmodernism and Truth." 1998. *Butterflies and Wheels* http://www.butterfliesandwheels.com/articleprint.php?num=13.

Dickens, Charles. *American Notes and Reprinted Pieces.* London: William Clowes and Sons, Limited, 1867–68.

Didion, Joan. "John Wayne: A Love Song." In Joan Didion, *Slouching towards Bethlehem,* 29–41. New York: Dell Publishing Company, 1969.

Dillard, Annie. *Pilgrim at Tinker Creek.* New York: Harper Collins, 1998.

Dodge, Martin, and Rob Kitchin, *Atlas of Cyberspace.* Harlow, England: Addison-Wesley, 2001.

Dutch, Robert A., ed. *The St. Martin's Roger's Thesaurus of English Words and Phrases.* New York: St. Martin's Press, 1965.

Eco, Umberto. *Travels in Hyperreality.* New York: Harvest/Harcourt Brace Jovanovich, 1990.

Eiseley, Loren. *The Invisible Pyramid.* New York: Charles Scribner's Sons, 1970.

Eisenberg, Anne. "The World through PC-Powered Glasses." *New York Times,* December 14, 2000.

Eliade, Mircea. *Myth and Reality.* Trans. Willard R. Trask. New York: Harper Colophon Editions, 1975.

Elshtain, Jean Lethke. "Two Cheers for Democracy." Review of Sheldon S. Wolin, *Tocqueville between Two Worlds. Washington Post,* November 11, 2001.

Emerson, Ralph Waldo. *Nature.* 1836. Boston: Beacon Press, 1991.

——. "Ode, Inscribed to William H. Channing." In *Early Poems of Ralph Waldo Emerson.* New York: Thomas Y. Crowell and Company, 1899.

Feder, Stuart. *The Life of Charles Ives.* Cambridge: Cambridge University Press, 1999.

Fitzgerald, F. Scott. *The Great Gatsby.* 1925. New York: Bantam Books, 1974.

Fixmer, Rob. "The Soul Amitai Etzioni, E-Communities Build New Ties, but Ties that Bind." *New York Times,* February 10, 2001.

Flores, Dan. *Horizontal Yellow: Nature and History in the Near Southwest.* Albuquerque: University of New Mexico Press, 1999.

——. *The Natural West: Environmental History in the Great Plains and Rocky Mountains.* Norman: University of Oklahoma Press, 2001.

Fortunato, Mary Beth. Review of Allen Mazur, *A Hazardous Inquiry. Electronic Green Journal* 12 (2000): 1–2. http://egj.lib.uidaho.edu/egj12/fortunato1.html.

Freyfogle, Eric T. *Bounded People, Boundless Lands: Envisioning a New Land Ethic.* Washington DC: Island Press, 1998.

Friedman, Ted. *Electric Dreams: Computers in American Culture.* New York: New York University Press, 2005.

——. "Semiotics of SimCity." *First Monday* (1999) 2. http://www.first monday.org/issues/issue4/friedman/index.html.

Garreau, Joel. "The Call of Beauty, Coming in Loud and Clear," *Washington Post,* February 19, 2002.

——. *The Nine Nations of North America.* Boston: Houghton Mifflin Company, 1981.

Gazaway, Rena. *The Longest Mile.* New York: Doubleday & Co., Inc., 1969.

Gelernter, David. "The Second Coming—A Manifesto." *Edge,* June 15, 2000. http://www.edge.org/documents/archive/edge70.html.

Gimblett, H. R., et al. "An Intelligent Agent Based Model for Simulating and Evaluating River Trip Scenarios along the Colorado River in Grand Canyon National Park." In H. R. Gimblett, ed., *Integrating GIS and Agent Based Modelling Techniques for Understanding Social and Ecological Processes,* 79–86. New York: Oxford University Press, 2000.

Goetzmann, William H. *Army Exploration in the American West; 1803–1863.* New Haven: Yale University Press, 1959.

——. *Exploration and Empire: The Explorer and the Scientist in the Winning of the American West.* New York: W. W. Norton & Company, 1966.

Gopnik, Adam. "The Unreal Thing." *New Yorker,* May 19, 2003, 68–72.

Gordon, Mary. "We Are All Spendthrifts Now." *New York Times,* June 6, 2000.

Green, Harvey. *The Uncertainty of Everyday Life, 1915–1945*. New York: HarperCollins, 1992.

Greene, Brian. "One Hundred Years of Uncertainty." *New York Times,* April 8, 2005.

Greenland, Colin. "A Nod to the Apocalypse: An Interview with William Gibson." *Foundation* 36 (Summer 1993): 12–15.

Greenwald, Maurine W., and Margo Anderson, eds. *Pittsburgh Surveyed: Social Science and Social Reform in the Early Twentieth Century.* Pittsburgh: University of Pittsburgh Press, 1996.

Greider, William. "One World of Consumers." In Rosenblatt, *Consuming Desires,* 23–36.

Grese, Robert E., and John R. Knott, eds. *Reimagining Place.* Special issue, *Michigan Quarterly Review* 40 (Winter 2001).

Grochowski, J. "Vegas' Virtual Reality: The City of Illusion Keeps Gambling on Reinvention." *Chicago Sun-Times,* October 8, 1995.

Hamilton, Andrew. "Theoretical Physics, in Video: A Thrill Ride to 'the Other Side of Infinity.'" *New York Times,* February 28, 2006.

Harries, Owen. "Understanding America." Lecture presented at Center for Independent Studies, Sydney, Australia, April 3, 2002. http://www.cis.org.au/occasional/harries030402.htm.

Hayes, Brian. *Infrastructure: A Field Guide to the Industrial Landscape.* New York: W. W. Norton and Company, 2005.

Hays, Samuel P. *Conservation and the Gospel of Efficiency.* Cambridge: Harvard University Press, 1959.

Heffernan, Virginia. "Climb In, Log On, Drop Out," *New York Times,* May 12, 2006.

Heilbroner, Robert L., and Aaron Singer. *The Economic Transformation of America: 1600 to the Present.* 2nd ed. New York: Harcourt Brace Jovanovich, 1984.

Heisenberg, Werner. *Physics and Philosophy: The Revolution in Modern Science.* New York: Harper and Brothers, 1958.

Heppenheimer, T. A. "The Rise of the Interstates: How America Built the Largest Network of Engineered Structures on Earth." *American Heritage of Invention and Technology* 7, no. 2 (1991): 8–18.

Hodgson, Godfrey. "The New Statesman Profile—Francis Fukuyama."

New Statesman Books, April 22, 2002. http://www.newstatesman.co.uk/
site.php3?new Template=NSArticle People&newDisplayURN=20020
4220011.

Horseman, Reginald. *The Frontier in the Formative Years, 1783–1815.* New
York: Holt, Reinhard and Winston, 1970.

Huizinga, Johan. *Homo Ludens: A Study of the Play Element in Culture.*
1938. Boston: Beacon Press, 1950.

Hyde, Anne Farrar. *An American Vision: Far Western Landscape and National Culture: 1820–1920.* New York: New York University Press, 1990.

Ives, Charles. *The Orchestral Music of Charles Ives.* Orchestra New England,
conducted by James Sinclair. Program notes and lyrics in foldout.
Westbury NY: Koch International, 3-7052-2, 1990.

———. *Songs.* Jan DeGaetani, mezzo-soprano, Gilbert Kalish, piano. Program notes and lyrics in foldout. New York: Warner Communications,
Nonesuch 71325-2, 1976.

———. *Three Places in New England.* Chicago Symphony, conducted by
Michael Tilson Thomas. Program notes 3–7. New York: Columbia
Masterworks CD MK 42382, 1988.

Jackson, J. B. "Jefferson, Thoreau and After: The Life and Death of
American Landscapes." *Landscape* 15 (Winter 1965–66): 25–27.

———. "The Order of a Landscape: Reason and Religion in Newtonian
America." In Meinig, *Interpretation of Ordinary Landscapes*, 231–32.

Jakle, John A. *The Tourist: Travel in Twentieth-Century North America.* Lincoln: University of Nebraska Press, 1985.

Jefferson, Thomas. "Jefferson to Maria Cosway." In Julian P. Boyd et al.,
eds., *The Papers of Thomas Jefferson*, vol. 10. Princeton: 1950.

Johnson, Reed. "Mass Media's Last Blast: I Want My MTV—and My TiVo,
Palm Pilot, iPod, Podcast and, of Course, Blog. So Does America Still
Have Any Interest in the Big, Lumbering, Predictable Media of Hol-
lywood and Manhattan?" *Los Angeles Times*, December 18, 2005.

Johnson, Steven. "Don't Fear the Digital." *Time Magazine*, March 27,
2006, 56.

Jonas, Hans. *The Phenomenon of Life: Toward a Philosophical Biology.* New
York: Delta Books, Dell Publishing Company, 1966.

Jones, Deborah. "Storybook Ending." *Vancouver Sun*, February 5, 2001.

Keller, Julia. "Plugged-in Proust: Has E-lit Come of Age?" *Chicago Tribune*, November 27, 2005.

Kellogg, Paul Underwood, ed. *The Pittsburgh District: Civic Frontage.* Vol. 5, *The Pittsburgh Survey of the Russell Sage Foundation.* New York: The Russell Sage Foundation, 1914. New York: Arno Press, 1974.

Kemmis, Daniel. "Learning to Think like a Region." *High Country News,* April 10, 2000, 7, 17.

Kidd, Dustin. *The 1930s.* Crossroads, American Studies at the University of Virginia. http://xroads.virginia.edu/~ma99/kidd/century/intro.html.

Kimball, Roger. "The Fortunes of Permanence." *New Criterion* 4, no. 21, Web special (2002): http://www.newcriterion.com/archive/20/summer02/fortunesofpermanence.htm.

———. *The Survival of Culture: Permanent Values in a Virtual Age.* London: Ivan R. Dee, 2002.

King, Clarence. *Mountaineering in the Sierra Nevada.* New York: W. W. Norton & Co., 1935.

Koppell, Jonathan G. S. "No 'There' Here." *Atlantic Monthly,* August 2000, 16–18.

Kundera, Milan. *The Book of Laughter and Forgetting.* New York: Penguin Books, 1981.

Kunkel, Benjamin. "The Deep End: A New Life of D. H. Lawrence." *New Yorker,* December 19, 2005, 92.

Lane, Belden C. *Landscapes of the Sacred: Geography and Narrative in American Spirituality.* Baltimore: Johns Hopkins University Press, 2001.

Lankford, John, ed. *Captain John Smith's America: Selections from His Writings.* New York: Harper Torchbooks, 1967.

Larkin, Jack. *The Reshaping of Everyday Life, 1790–1840.* New York: Harper and Row, 1988.

Lavadie, Paul G. "USA Not as Good as We Hoped, nor as Bad as We Feared." *USA Today,* October 26, 1999.

Lehmann, Chris. "Hit the Road to Dreamland." *DeepRead,* 2001. http://www.feedmag.com/deepread/dr41110fi.html.

Leland, John. "Family's Choices Can Blunt the Effect of Video Violence." *New York Times,* September 25, 2000.

———. "Pixel Canvas; The Gamer as Artiste." *New York Times,* December 4, 2005.

Lemons, John. "National Park Management and Values." 1982.

Lennon, John, and Paul McCartney. "Nowhere Man." 1964. *Beatles/ 1962–1966*. http://www.thebeatles.com.

Leopold, Aldo. *A Sand County Almanac, and Sketches Here and There*. 1949. New York: Oxford University Press, 1989.

Lessig, Lawrence. "The Internet under Siege," *Foreign Policy* (November– December 2001): 1–12.

Lewis, Peirce. "Axioms for Reading the Landscape." In Meinig, *Interpretation of Ordinary Landscapes*, 11–32.

Lewis, Tom. *Divided Highways: Building the Interstate Highways, Transforming American Life*. New York: Viking Penguin, 1997.

Light, Andrew, and Jonathan M. Smith, eds. *Philosophies of Place*. Lanham: Rowan and Littlefield Publishers, Inc., 1998.

Lindgren, Hugo. "Generation Xbox." *New York Times Book Review*, December 18, 2005.

Lohr, Steve. "Welcome to the Internet, the First Global Colony." *New York Times*, January 9, 2000.

Lukacs, John. "It's the End of the Modern Age." *Chronicle of Higher Education*, April 26, 2002. http://chronicle.com/weekly/v48/i33/ 33b00701.htm.

Lundberg, Isabel Cary. "World Revolution, American Plan." *Harper's Magazine*, December 1948, 38–46.

Lyman, Rick. "Virtual Reality Comes Back in New Guise: Collaboration." *New York Times*, July 31, 2000.

"Major Land Resource Area (MLRA) Boundaries." Agricultural Handbook 296. Washington DC: USDA, Soil Conservation Service, 1981.

Mandoki, Katya. "Sites of Symbolic Density; A Relativistic Approach to Experienced Space." In Light and Smith, *Philosophies of Place*, 73–96.

Maplas, Jeff. "Finding Place: Spatiality, Locality, and Subjectivity." In Light and Smith, *Philosophies of Place*, 22–23.

Matthiessen, Peter. *The Snow Leopard*. New York: Viking Press, 1978.

Mayda, Chris. "Geographing.com: Guiding Principles." July 2000.

Mazur, Allen. *A Hazardous Inquiry: The Rashomon Effect at Love Canal*. Cambridge: Harvard University Press, 1998.

McCullough, David. *The Great Bridge: The Epic Story of the Building of the Brooklyn Bridge*. New York: Simon and Schuster, 1972.

McHarg, Ian. *Design with Nature*. Garden City NY: Doubleday & Company, 1971.

McShine, Kynaston. *The Natural Paradise: Painting in America, 1800–1950.* New York: Museum of Modern Art, 1976.

Meinig, Donald W. "The Beholding Eye." In Meinig, *Interpretation of Ordinary Landscapes,* 33–50.

———, ed. *The Interpretation of Ordinary Landscapes: Geographical Essays.* New York: Oxford University Press, 1979.

Melville, Herman. *Moby Dick; or, The Whale.* New York: Modern Library, 2000.

Merchant, Carolyn. *Reinventing Eden: The Fate of Nature in Western Culture.* New York: Routledge, 2003.

Merk, Frederick. *Manifest Destiny and Mission in American History: A Reinterpretation.* New York: Vintage Books, 1966.

Miller, Angela. *The Empire of the Eye: Landscape Representation and American Cultural Politics, 1825–1875.* Ithaca: Cornell University Press, 1993.

Miller, Donald L. *City of the Century: The Epic of Chicago and the Making of America.* New York: Touchstone/Simon & Schuster, 1996.

Miller, Perry. *Errand into the Wilderness.* Cambridge: Belknap Press of Harvard University Press, 1956.

———. *Nature's Nation.* Cambridge: Belknap Press of Harvard University Press, 1967.

Mitchell, John Hanson. *Ceremonial Time.* New York: Warner Books, 1986.

Morris, Wright. "To the Woods." In Owen Thomas, ed., *Henry David Thoreau: Walden and Civil Disobedience,* 384–89. New York: W. W. Norton and Company, 1966.

Muir, John. *Our National Parks.* Boston: Houghton Mifflin, 1901.

———. "The Wild Parks and Forest Reservations of the West." *Atlantic Monthly,* January 1898, 26–28.

"The Multitasking Generation." *Time Magazine,* March 27, 2006, 53.

Norton, Bryan, and Bruce Hannon. "Democracy and Sense of Place Values in Environmental Policy." In Light and Smith, *Philosophies of Place,* 119–20.

Novak, Barbara. *Nature and Culture: American Landscape and Painting, 1825–1875.* New York: Oxford University Press, 1980.

O'Connell, Pamela LiCalzi. "Mining the Minds of the Masses." *New York Times,* March 8, 2001.

O'Keeffe, Georgia. *Georgia O'Keeffe*. New York: Penguin Books, 1976.

Opie, John. "Energy and the Rise of American Industrial Society." In Ian Barbour, Harvey Brooks, Sanford Lakoff, and John Opie, eds., *Energy and American Values*, 1–23. New York: Praeger Special Studies, 1982.

———. *Nature's Nation: An Environmental History of the United States*. Fort Worth: Harcourt Brace and Company, 1998.

———. "Seeing Desert as Wilderness and as Landscape—An Exercise in Visual Thinking." In "Proceedings of Our National Landscape: A Conference on Applied Techniques for Analysis and Management of the Visual Resource, April 23–25, 1979, Incline Village, Nevada," 101–8. General Technical Report PSW-35. Berkeley: USFS, USDA, Pacific Southwest Forest and Range Experiment Station, 1979.

Otto, Rudolph. *The Idea of the Holy*. 2nd ed. New York: Oxford University Press, 1923.

Overbye, Dennis. "Quantum Trickery: Testing Einstein's Strangest Theory." *New York Times*, December 27, 2005.

Parkes, Henry Bamford. *The American Experience: An Interpretation of the History and Civilization of the American People*. New York: Vintage Books, 1959.

Pascal, Blaise. *Pensées*. Trans. A. J. Kraisheimer. Baltimore: Penguin Books, 1966.

Percy, Walker. *The Message in the Bottle: How Queer Man Is, How Queer Language Is, and What One Has to Do with the Other*. New York: Picador/ Farrar, Straus and Giroux, 1954–75.

Persons, Stow. *American Minds: A History of Ideas*. New York: Henry Holt and Company, 1958.

Petroski, Henry. *Engineers of Dreams: Great Bridge Builders and the Spanning of America*. New York: Random House, 1995.

Pomeroy, Earl. *In Search of the Golden West: The Tourist in Western America*. New York: Alfred A. Knopf, 1957.

Potter, David M. *People of Plenty: Economic Abundance and the American Character*. Chicago: University of Chicago Press, 1954.

Pyle, Ernie. *Here is Your War: Story of G. I. Joe*. Cleveland: World Publishing Company, 1943.

Pyne, Steve. *How the Canyon Became Grand: A Short History*. New York: Penguin Books, 1998.

Radosh, Daniel. "Cyber City." *New Yorker*, December 19, 2005, 36, 39.

Ritzer, George, and Allan Liska. "'McDisneyization' and 'Post-Tourism': Complementary Perspectives on Contemporary Tourism." In Chris Rojek and John Urry, eds., *Touring Cultures: Transformations of Travel and Theory*, 96–99. London: Routledge, 1997.

Rosenblatt, Roger. *Consuming Desires: Consumption, Culture, and the Pursuit of Happiness*. Washington DC: Island Press, 1999.

Rothstein, Mervyn. "Ingmar Bergman, Master Filmmaker, Dies at 89." *New York Times*, July 31, 2007.

Rousseau, Jean-Jacques. *The Social Contract and Discourses*. Trans. G. D. H. Cole. Rev. J. H. Brumfitt and John C. Hall. Updated by P. D. Jimack. London: J. M. Dent, 1993.

Runte, Alfred. *National Parks: The American Experience*. 2nd. ed., rev. Lincoln: University of Nebraska Press, 1987.

Rush, Michael. "In Love with Reality Truly, Madly, Virtually." *New York Times*, January 8, 2006.

Rydell, Robert W. *All the World's a Fair*. Chicago: University of Chicago Press, 1984.

Sacks, Oliver. *An Anthropologist on Mars*. New York: Vintage Books/Random House, 1995.

———. "The Mind's Eye: A Neurologist's Notebook." *New Yorker*, July 28, 2003, 20, 46–59.

Sagoff, Mark. *The Ethics of Consumption*. College Park MD: Institute for Philosophy and Public Policy, 1995.

Sanders, Scott Russell. *The Country of Language*. Minneapolis: Milkweed Editions, 1999.

———. *Secrets of the Universe: Scenes from the Journey Home*. Boston: Beacon Press, 1991.

———. *Staying Put: Making a Home in a Restless World*. Boston: Beacon Press, 1993.

Schaub, George. "Eyes Wide Open: When the Travel and Photo Bug Conspire." *Shutterbug*, May 2005, 27.

Schiesel, Seth. "Video Games Are Their Major, So Don't Call Them Slackers." *New York Times*, November 22, 2005.

———. "Welcome to the New Dollhouse." *New York Times*, May 7, 2006.

Schnell, Izhak. "Transformations in the Myth of the Inner Valleys as a Zionist Place." In Light and Smith, *Philosophies of Place*, 97.

Schor, Juliet. "What's Wrong with Consumer Society? Competitive Spending and the 'New Consumerism.'" In Rosenblatt. *Consuming Desires*, 37–50.

Schulman, Edmund. "Bristlecone Pine, Oldest Known Living Thing." *National Geographic* 113 (1958): 354–72.

"Seven Questions: Battling for the Internet." *Foreign Policy* (November 2005): http://www.foreignpolicy.com/story/cms.php?story_id=3306.

Shepard, Paul. "Virtually Hunting Reality in the Forests of Simulacra." In Michael E. Soule and Gary Lease, eds., *Reinventing Nature? Responses to Postmodern Deconstruction*, 21–25. Washington DC: Island Press, 1995.

Smith, Jonathan, Andrew Light, and David Roberts. "Introduction: Philosophies and Geographies of Place." In Light and Smith, *Philosophies of Place*, 1–20.

Sobel, David. "The Sky Above, the Internet Below." *Sanctuary: The Journal of the Massachusetts Audubon Society* 40 (January–February 2001): 12–18.

Sopher, David E. "The Landscape of Home: Myth, Experience, Social Meaning." In Meinig, *Interpretation of Ordinary Landscapes*, 129–52.

Steinberg, Ted, *Slide Mountain; or, The Folly of Owning Nature*. Berkeley: University of California Press, 1995.

Stephanson, Anders. *Manifest Destiny: American Expansion and the Empire of Right*. New York: Hill and Wang, 1995.

Stevens, William K. "Latest Threat to Yellowstone: Admirers Are Living It to Death." *New York Times*, September 15, 1994.

Stilgoe, John R. *Outside Lies Magic: Regaining History and Awareness in Everyday Places*. New York: Walker and Company, 1998.

Tall, Deborah. *From Where We Stand: Recovering a Sense of Place*. Baltimore: Johns Hopkins University Press, 1993.

Terry, Sara. "A Supper's Experiment: Can She Really 'Eat Locally'?" *Christian Science Monitor*, May 14, 2003. http://www.csmonitor.com/2003/0514/p14s02-lifo.html.

Thomas, Bob. *Walt Disney: An American Original*. New York: Hyperion, 1996.

Thompson, Clive. "Game Theories." *Walrus Magazine*, May 6, 2004. http://walrusmagazine.com/article.pl?sid-04/05/06/1929205.

Thoreau, Henry David. *The Maine Woods*. Princeton: Princeton University Press, 2004.

Tobin, James. *Ernie Pyle's War: America's Eyewitness to World War II*. New York: Free Press, 2006.

Tocqueville, Alexis de. *Democracy in America*. Trans. George Lawrence. Ed. J. P. Mayer. New York: Doubleday Anchor, 1969.

Tuan, Yi-Fu. "The Desert and I: A Study in Affinity." In Grese and Knott, *Reimagining Place*, 7–16.

———. *Escapism*. Baltimore: Johns Hopkins University Press, 2000.

———. *Space and Place: The Perspective of Experience*. Minneapolis: University of Minnesota Press, 1977.

Turkle, Sherry. *Life in the Screen: Identity in the Age of the Internet*. New York: Simon and Schuster, 1995.

Turner, Frederick. *Spirit of Place: The Making of an American Literary Landscape*. Washington DC: Island Press, 1989.

Tuveson, Ernest Lee. *Redeemer Nation: The Idea of America's Millennial Role*. Chicago: University of Chicago Press, 1968.

Warner, Sam Bass. *The Urban Wilderness: A History of the American City*. Berkeley: University of California Press, 1995.

Weinberg, Steven. "Five and a Half Utopias." *Atlantic Monthly*, January 2000, 114–16.

Weisberger, Bernard A. *The New Industrial Society*. New York: John Wiley and Sons, 1969.

Wertheim, Margaret. *The Pearly Gates of Cyberspace: A History of Space from Dante to the Internet*. New York: W. W. Norton, 1999.

White, Lynn, Jr. *Dynamo and Virgin Reconsidered: Essays in the Dynamism of Western Culture*. Cambridge: MIT Press, 1968.

"Will the Internet Change Humanity?" Symposium on television series *Closer to Truth*, May 2000. http://www.closertotruth.com/topics/technologysociety/102/102transcript.html.

Wilson, Eric. *Romantic Turbulence: Chaos, Ecology, and American Space*. New York: St. Martin's Press, 2000.

Wines, Michael. "Russia Finds Virtue in a U.S. Victory, Looking toward Integration with West." *New York Times*, May 14, 2002.

Winther, Oscar Osburn. *The Transportation Frontier: Trans-Mississippi West, 1865–1890*. New York: Holt, Reinhart and Winston, 1964.

Wolcott, Jennifer. "In Search of the Ripe Stuff." *Christian Science Monitor*, May 14, 2003. http://www.csmonitor.com/2003/0514/p14s03-lifo.html.

Wolf, Joshua. "Hidden Kingdom—Disney's Political Blueprint." *American Prospect*, March 21, 1995. http://www.waltopia.com/hiddenkingdom.html.

Wolfe, Linnie Marsh. *Son of the Wilderness: The Life of John Muir*. Madison: University of Wisconsin Press, 1945.

"Worlds without End." Review of Edward Castronova, *Synthetic Worlds: The Business and Culture of Online Games*. *Economist*, December 14, 2005, 42–43.

Worster, Donald. *A River Running West: The Life of John Wesley Powell*. New York: Oxford University Press, 2001.

Wright, John Kirkland. *Human Nature in Geography: Fourteen Essays, 1925–1965*. Cambridge: Harvard University Press, 1966.

Wright, Thomas. "What's for Afters?" *Arts.telegraph.co.uk*, April 6, 2002. http://www.arts.telegraph.co.uk/arts/main.jhtml?xml=/arts/2002/04/01/bohea31.

Wright, Will. "Dream Machines . . . How Games Are Unleashing the Human Imagination." *Wired*, April 2006. http://www.wired.com/wired/archive/14.04/wright_pr.html.

Zeller, Tom, Jr. "A Generation Serves Notice: It's a Moving Target." *New York Times*, January 22, 2006.

Zimmerman, Warren. *First Great Triumph: How Five Americans Made Their Country a World Power*. New York: Farrar, Straus and Giroux, 2002.

Index

Page numbers in italic refer to illustrations

Abbey, Edward, on consumerism, 123–24

Abert, John J., *117*

Absalom, Absalom (Faulkner), 162

Activision, 96

Adams, Ansel, 4, *5*

Adams, Henry, 86

Adirondack Mountains, 49, *51*

adventure vs. ecotourism, 81

Agee, James, 151–52, *168*

Age of Progress, 112–13

Age of Reason, 125

Albanese, Cathy, 24

Albany, 56

Albert, Prince, 84

Alberta, West Edmonton Mall in, *102–3*

Alison, Archibald, 59

Allen, Woody, *The Purple Rose of Cairo*, 6

Altamira, 22

America: American perceptions of, 45–49; European perceptions of, 43–44; global perceptions of, 144–45; invention of, 42–43; as Kingdom of God, 140–41; rural landscapes of, 49–52; virtual reality of, 23, 29, 30

American Empire, 137

American Revolution, 170

Amritsar, 169

Anderson, Sherwood, 162

Anthony, E., and H. T., *Yosemite Valley, California*, 65

antiquity, 72–73

Appalachia, 51, 149
Appleton's Illustrated Handbook of American Travel, 57
Arcadia, 49–50
archaeology, as metaphor, xi–xii
architecture, 19, 78, 94; of Columbian Exposition, 92–93
Arena Chapel (Padua), 22
Arendt, Hannah, 38
artists, art, 19, 59; and desert landscapes, 75–76, 79; landscape, 52–54, 61–66, 65; and the sublime, 66–69
Ash, Timothy Garton, 144
Atlanta, world's fair in, 88
atomic bomb, 183
Augustine, St., *Confessions*, 200
Auschwitz, 169
Authentic America, xiv, 50
authenticity, 46
autism, 31
automobile travel, 61, 121

Babi Yar, 170
Bacon, Francis, 25
Badgley, Catherine, 200–201
Bailey, Dana K., 180
Bailey, Robert G., 188
Banvard, John, *Panorama of the Mississippi River*, 62
Battle Creek, 48
"The Battle Hymn of the Republic," 139
Bazelon, David T., 129
"The Bear" (Faulkner), 162
Beatles, "Nowhere Man," 177
Beecher, Henry Ward, 62

Bellamy, Edward, *Looking Backward*, 93
Bell Telephone, 94
Berg, Peter, 190
Bergman, Ingmar, 177
Berners-Lee, Tim, 8
Berry, Wendell, 198, 205
Beveridge, Albert J., 140
Bhatia, Sabeer, 8
Bierstadt, Albert, 1, 21, 52, 54, 60, 65
Bingham, George Caleb, *The Emigration of Daniel Boone*, 133, *133*
bioregionalism, 190
blindness and mental mapping, 26–28
Boorstin, Daniel J., 59, 111, 170, 173, 205
Borgmann, Albert, 8
Boston, eating locally in, 132–33
Botton, Alain de, 66
boys and video games, 14–15
Brand, Stewart, 137
Bread & Circus, 134
Breton, André, 22
Bridal Veil Falls, 61
bristlecone pines, 178–79; symbolism of, 180–85
Bristlecone Pines (Fiddler), *178*
Brodsky, Joseph, 182; on autonomy of nature, 179–80
Brooklyn, 108, *109*, 110
Brooklyn Bridge, 108, *109*
Brown, Capability, 70
Brown, Dona, 51
Brown, Michael, 30

Buffalo, Pan-American Exposition in, 88, 90
Burke, Edmund, 66, 200
Burned-Over District, 153
Bushnell, Horace, 49–50, 139

Cable, George Washington, 162
Calhoun, John C., 137
California, 72, 114, 137, 153; eating locally in, 131–32
California Gold Rush, panoramas of, 62
Callicott, Baird, 195
Campbell, Joseph, 204
camping, 49, 80
The Camp Meeting (Ives), 167
Canyonlands, 75, 76
capitalism, 123, 128, 138
Carnegie, Andrew, 128
Casey, Edward S., on place, 151, 152–53, 201, 206
Cass, Lewis, 137
Castronova, Edward, 13, 17
Cather, Willa, 51; *Death Comes for the Archbishop*, 192; "Paul's Case," 105
Catskill Mountain House, 57
Catskill Mountains, 57, 70
cave paintings, 22
CCC. *See* Civilian Conservation Corps (CCC)
Central Park, 47
Central Valley, 114
Century of Progress, 98–99
CERN. *See* European Organization for Nuclear Research (CERN)
Chace, James, 141

Chartres Cathedral, 22
Chicago, 48; and Century of Progress, 98–99; and World Columbian Exposition, 87, 88, 90, 91, 92–94
childhood: home place during, 158–60; memory of, 160–61
childishness, 29, 35
children and digital games, 13–16
Choate, Rufus, 52
Christianity, apocalyptic warfare and, 139–40
Church, Frederick, 52, 54
cities, 114–15, 121; idealized, 95–96; liveable, 122–23
City of God, creating the, 21–22
city parks, 70
Civilian Conservation Corps (CCC), 49
Civilization, 14–15, 96
Civil War battlefields, 47
Claude, George, 61
Clermont, 117
Coal Mountain, 48
Cobb, Irvin, 32
Cobb, John, 130
cognitive mapping, 26–27
Cohen, Michael P., on bristlecone pines, 179, 180–81, 182–84
Cole, Thomas, 21, 54, 75; *The Course of Empire*, 93; *Falls of the Kaaterskill*, 67; *The Oxbow*, 52, 53
Collier, David C., 92
Colorado, 71–72

Columbus, Christopher, x–xi
commons, cyberspace as, 8–9
Confessions (St. Augustine), 200
Connecticut Oxbow, 52, *53*
Connecticut River, 55
Conron, John, 32
consciousness, alternative to, 29
conspicuous consumption,
 129–30
Constitution, 170
Consumer America, xii, 108, 123
consumerism, 123–24, 142; food
 and, 125–26; global hegemony
 and, 143–44; Tocqueville on,
 126–27; Twain on, 127–28;
 Veblen on, 129–30
Consummation of Empire (Cole), 93
Cook, Jay, 71
Corliss, George Henry, 85, 86
corporations, multinational, 145
Cortez, Hernando, 23
Cosway, Maria, 69
counties, map of, *184*
The Course of Empire (Cole), 93
creation stories, Grand Canyon
 and, 78
Crèvecoeur, Michel-Guillaume-
 Jean de, 49, 117
Crofutt, George A., 135
Crowley, John, 99
Cruz-Neira, Caroline, 19
Crystal Palace, 84, 85
Cumberland Gap, *133*
Currier & Ives: cityscape by, *109*;
 rural landscapes by, 2, *172*;
 technology themes by, *116*

Curtis, G. E., 63
cyberspace, 12, 23, 27, 28, 90,
 199; as commons, 8–9; home
 and, 7–8; perceptions of,
 6–7; place in, 156–57; vs. real
 world, 9–10; virtual reality in,
 ix, xii, xiii–xiv, 4–6, 19, 37–38;
 wealth in, 145–46
Czolgosz, Leon, 90

Daly, Herman, 130
Danbury, 165
Dante, *Divine Comedy*, 22
Davis, Richard Harding, 92
Death Comes for the Archbishop
 (Cather), 192
deception, virtual reality as, 33
DeFanti, Tom, 19
Democracy in America (Tocque-
 ville), 126–27
dendrochronology, 178–79
Dennett, Daniel, 195
deserts, 73, *74*; artists' views of,
 75–76; as sacred space, 76–79
Destruction and Desolation (Cole),
 93
Detroit tourism, 48
Dewey, John, 151
Dickens, Charles, 26, 43–44
Dillard, Annie, xvi, 32, 35, *175*;
 Pilgrim at Tinker Creek, 201–2
Disney, Walt, on EPCOT, 100–102
Disney enterprises, 46, 99
Disneyland Effect, 6
Disneyworld, 88, *89*
Divine Comedy (Dante), 22
Dodge, Grenville M., 118,

Dodge, Martin, 12, 156, 157
Donne, John, 52
Dostoyevsky, Fyodor, 35
Drakulich, Mike, 181–82
Dubai, 103
Dubos, Rene, *The Wooing of the Earth*, 198
Durrell, Lawrence, 176
Dust Bowl, 21
Dutton, Clarence Edward, 77, 78, 79

Eagleton, Terry, 181–82
eating locally, 131–32, 134
Ebejer, Simon, 96
Eco, Umberto, 42, 183
ecology, 194, 198, 205, 220–21n43; private property, 196–97
ecoregions, 186–87, *186*, *187*, 190
ecosystems, 7, 41, 43, 81, 134
ecotourism, 80–81
Edison, Thomas, 84, 93
Edmonton, shopping mall in, 102–3
Einstein, Albert, xi, xviii
Eiseley, Loren, xvi, 169
Electric Age, 93
electric power, 95; at Columbian Exposition, *84*, 93–94
Elements of Political Economy (Wayland), 127
elevators, 94
Eliade, Mircea, 31, 32, 45, 153, 174; on sacred places, 203, 204

Eliot, T. S., 37, 161; *Wasteland*, 177
elites, engineers as, 109
Emerson, Ralph Waldo, 23–24, 25, 37, 45, 54, 80, 124
The Emigration of Daniel Boone (Bingham), *133*
Engineered America, xii, 108, 111–12
engineers, engineering: Age of Progress and, 112–13; as heroes, 109–12; and infrastructure, 114–15; and nature, 108–9
environment, 21, 41, 43, 114, 138; and ecotourism, 80–81; virtual, 19–20
Environmental Protection Agency, 188
environmental science, 205–6
EPCOT. *See* Experimental Prototype Community of Tomorrow (EPCOT)
epidemics: typhoid, 122; urban, 169
epiphanies, 35
Eskimos, xiii
ethics: toward land, 193–94; virtual world, 36
Europe, and invention of America, 42–43
European Organization for Nuclear Research (CERN), 8
Everglades, 190
EverQuest, 13, 17, 18
exceptionalism, American, 138–39

Experimental Prototype Community of Tomorrow (EPCOT), 88, 99, 100–102

expositions. See world's fairs

factories, 48, 122

fairs, 48, 87–88. See also world's fairs

Falls of the Kaaterskill (Cole), 67

fantasy, 30; fairs as, 87–88, 91

farms, farmers, 2; and tourism, 49–50

Farrow, Mia, in *Purple Rose of Cairo*, 6

Faulkner, William, 162

Feder, Stuart, 165

Ferguson, Melville, 72

Fiddler, Claude, *Bristlecone Pines*, *178*

Finger, Charles, 32, 72

First Nature, xv, 7, 95, 107, 111, 126, 129, 131; autonomy of, 179–80; transformation of, 136–37

First Place, 153–54

Fitzgerald, F. Scott, 42

Flores, Dan, 190, 200

Florida, 137, 190; EPCOT in, 88, 101

Flushing Meadows, 99

food, availability of, 125–26, 131–32, 134

Ford factories, tourism at, 48

Fort Clatsop, 47

Foucault, Michel, 150, 177–78

Fourteen Points, 134

Frank, Robert, 130

Frémont, John C., 75, 179

Freud, Sigmund, xii

Freyfogle, Eric, on private property, 196–98

Friedman, Ted, 95

frontier, 128; Turner's definition of, 140, 172

Fukuyama, Francis, 37, 144

Fulton, Robert, 117

Gage, Lyman, 93

games: computer and video, 12–13, 14–15, 18; as magical experience, 16–17; *The Sims*, 13–14

Garden City movement, 49

Garden of Eden, 21

Garden of the Gods, 57–58

garden parks, 70

Garreau, Joel, *Nine Nations of North America*, 51–52

Gary IN, 48

Gast, John, *Manifest Destiny*, *135*

GCRTSIM. See *Grand Canyon River Trip Simulator* (GCRTsim)

gender, and video games, 14–15

genius, 35

geographic systems, natural, 186–87

geography, xii, 9, 37, 107, 170

geopiety, 203

Gettysburg battlefield, 47

Ghermezian brothers, 102, 103

ghost towns, 47

Gibbs, Lois, 30

Gibson, William, 12;
    Neuromancer, 6
Gilpin, William, 69–70
Giotto, Arena Chapel, 22
girls and *Sims* games, 14
Glacier National Park, 58
global village, 177
Golden Gate Bridge, 47
Good Neighbor policy, 141
Gopnik, Adam, 33–34
Gordon, Mary, 124–25
government, state vs. federal, 115
Gowan, Peter, 145
Graceland, 47
Grand Canyon, 28, 32, 49, 54,
    58, 71, 77; as iconic place,
    47–48; as sacred place, 78–79
Grand Canyon River Trip Simulator
    (GCRTsim), 17–18
Grandin, Temple, 31
Grand Teton(s), 4, 5, 59
Grand Tour, 55; as virtual reality,
    56–58
Grant, U. S., 86
Gray, Elisha, 93
Great American Desert, 75
Great American Outdoors, 49
Great Basin, 179, 183
Great Depression, 49
Greater Yellowstone Ecosystem,
    188–90
Great Lakes Region, 190
Great Outdoors, mythical, 1–2
Great Plains, 21, 51, 59, 119
Green, Harvey, 124
Guernica, 169–70
guidebooks, imagery in, 56–58

Hahn, William, *Yosemite Valley*, 60
Haines Corners, 57
halls of fame, 46, 47
Hamilton, Andrew J. S., xi
Hannibal MO, Twain's imagery of,
    162–64
Harries, Owen, 144–45
Harris, Samuel, 139
Harvey, David, 203
Hay, John, 141
Hayden, Ferdinand Vandiveer, 79
health, 76, 122
health spas, 56
Heath, Don, 146
Heavenly City, 21–22
Hegel, Friedrick, 110
hegemony, American, 105,
    143–44
Heidegger, Martin, 153, 155, 199
Heilbroner, Robert, 129
Heinz, 48
Heisenberg, Werner, xviii, 176
Hemingway, Ernest, 51
Hering, Rudolph, 110–11
heroes, engineers as, 109–11
Herrick, Robert, 94
Hershey factory, 46
hierophany, 32, 204
highways, 121
hiking, 49
Histoire de Juliette (Sade), 22
historic sites, national identity
    and, 47
history, 37, 45, 138; place in,
    192–95; virtual reality of, 20–21
Hodgson, Godfrey, 144

Hoffer, Eric, 171–72
Hollander, Norman, 90–91
Hollywood Boulevard, 47
Holmes, Oliver Wendell, 64–65
home, cyberspace at, 7–8, 10
home entertainment industry, 7–8
home place, x, xvii, 107, 131, 134, 176; inhabiting, 192–93; mapping, 158–60; in mobile society, 171–72; orientation by, 154–55; power of, 161–62; story of, 162–63
Homo ludens (Huizinga), 16
Hooker Chemical Company, 30
Hoover, Herbert, 115
hotels, 71–72
"The Housatonic at Stockbridge" (Ives), 166
How the Canyon Became Grand (Pyne), 78
Hudson River School, 67
Hudson Valley, 46, 47, 55, 56
Huizinga, Johan, on play, 16–17
humanism, Renaissance, 43
Huxtable, Ada Louise, 102
Hyde, Anne Farrar, 62
Hyde Park (New York), 47
hydroelectric power, Niagara Falls and, 108–9
hydrological units, mapping by, 185
hyperreality, 42
Hyun, Young, 9

icons, American, 47
identity: American, xii, 47, 52,

53–54, 203; New World, 42, 43
IGE, 17
IMAX, 21
imperialism, European, 42
Indian Wars, 72
individualism, 18, 138, 198
Industrial Revolution, 125
industry, industrialization, 84, 129; American, 48–49, 85; and pollution, 168, 169
The Influence of Sea Power on History (Mahan), 141
infrastructure, 7, 8, 114–15, 119, 169; urban, 120, 121, 122–23
Ingalls, John J., 93
Internet, 8, 10, 12, 13, 37, 42, 90, 146
invention(s), nature and, 113–14
Iowa, 47
ironworks, 48
Ives, Charles, and music of place, 164–68
Ives, Joseph Christmas, 78

Jackson, John Brinkerhoff, 29, 155, 175, 177
Jackson, William H., 3
James, William, 128
James Brothers, 72
Jefferson, Thomas, 49, 69, 86
Jenkins, Henry, 19
Jonas, Hans, 202
Jones' Pantoscope of California, 62
journeys, photography and, 40–41
Jung, Carl, 204
Just v. Narinette, 196

Kaaterskill Railway, 57 (fig.)
Kansas, 50
Kellogg, D. Otis, 87
Kellogg's, 48
Kelvin, Lord, 108
Kemmis, Daniel, 190
Kennedy Space Center, 47
Keyserling, Count, 50
Kidd, Dustin, 90
Kimball, Roger, 36
Kinetoscope, 93
King, Clarence, 69, 79
Kingdom of God, America as,
    140–41
Kitchin, Rob, 12, 156, 157
knowledge, 200
Knoxville TN, 88, 151
Kodachrome, 63
Koppell, Jonathan, 23
Kubrick, Stanley, *2001: A Space
    Odyssey*, 33
Kuhn, Thomas, xviii, 184
Kundera, Milan, 178

LaMarche, Valmore C., Jr., 183
Lambert, Darwin, 184
land ethic, 193–94, 205
"The Land Ethic" (Leopold),
    193–94
landscapes, 21, 32, 55, 59, *60,*
    112, 201; American, 44,
    45–46; art of, 52–54, 61–66;
    desert, 73–79; infrastructure
    in, 114–15; interchangeable,
    174–75; as invented spaces,
    29–30; picturesque, 69–70;

rural, 2, 49–52; scenic beauty
    of, 194–95; as sublime, 66–69;
    Western, 71–73
land surveys of the West, 79–80
Lane, Belden C., 204
Langford, Nathaniel P., 79
Lascaux, 22
Last Nature, xiii, 108
Las Vegas, 46, 183
League of Nations, 141
Lebow, Victor, 125
Leland, John, 7
Leopold, Aldo, 35, 45, 201, 205;
    land ethic of, 193–94
Levittown, 173
Life on the Mississippi (Twain), 164
Lindgren, Hugo, 6–7, 13
Link, Edwin, 36
literature, 19, 54, 56
living room and cyberspace, 7–8
Locke, John, 111
Lodge, Henry Cabot, 141
lodges, 58
Lohr, Steve, 146
London world's fair, 83–86
Looking for Water, 20
Loop, 94
Los Angeles, eating locally in,
    131–32
Louisiana Purchase, 137
Louisiana Purchase Exposition,
    92
Love Canal, 30–31
Lowell MA, 48
Lowenthal, David, 80
Lukacs, John, 145–46

Lummis, Charles F., 92
Lundberg, Isabel Cary, 142
Lusseyran, Jacques, 26–27
Lynch, Kevin, 190

Machine Age, 90, 98
magical realism, 19
Magic Kingdom, 99
magic lantern shows, 62–64
Mahan, Alfred Thayer, *The Influence of Sea Power on History*, 141
Maine, 54
Main Street, Disneyland's, 36, 37
"Major Land Resource Area (MLRA) Boundaries," 188
Mall of America, 88, 103
malls, 102–3
Mandoki, Kayta, 169–70, 199
Manhattan, 96
Manifest Destiny, 3, 66, 105, 118, 134, 135(fig.), 136–38, 141; world's fairs and, 83, 87, 91–92
Manitou Springs Onaledge, 58(fig.)
mapping, maps, *184*; home place 158–61; mental, xiv, 26–28; physiographic, 55, *185*, *186*, 187–88
Marin, John, desert landscapes by, 75–76
marketplaces, 17, 196
Marshall Plan, 142
Martha's Vineyard, 50
Massachusetts, 50, 152
mass media, 95

material goods, consumerism and, 124, 131
materialism, 126–27
The Matrix, 33, 34
Matthiessen, Peter, 176
Mauch Chunk, 48
Mazur, Allan, 30–31
McCullough, David, 110
McHarg, Ian, 169
McKinley, William, 90, 141
McLuhan, Marshall, 177
mechanical panoramas, 61–62
Meining, Donald, 150, 151, 155, 205
Melville, Herman, 127
memories, 41; of home place, 159–60; and reality, 160–61
Merchant, Carolyn, 21
Merk, Frederick, 134
Merleau-Ponty, Maurice, 199, 202
metropolises, virtual, 13
metropolitan areas, *187*
Meyer, David E., 13
Michigan, 51, 137
migration, 172–74
military, American superiority, 138
Millennials, 12–13
Miller, Angela, 54, 59
Miller, Donald L., 93
Miller, Perry, 154
Mills, Robert, 118
Minneapolis, Mall of America in, 88, 103
missionaries, Christian, 42

Mississippi Delta, 47
Mississippi River, 51, 164
Mitchell, John Hanson, 152
mobility: American, 172–74; and
    loyalty to place, 171–72
Moctezuma, 23
Mohonk Mountain House, 70
Mondrian, Piet, 20
money in virtual world, 145–46
Monument Valley, 78
moral authority, American, 134,
    136
Moran, Thomas, 21, 52, 54, 79;
    The Mountain of the Holy Cross,
    1, 3
Mormons, 153
Morrell, Daniel J., 86
Morris, Wright, 56
Morse, Samuel F. B., 134
The Mountain of the Holy Cross
    (Moran), 1, 3 (fig.)
Mount Rainier National Park, 81
Mount Rushmore, 47, 72
Mount Vernon, 47
movement, freedom of, 171–72,
    173
movie lectures, 63
Moving Mirror of the Overland Trail
    (Wilkins), 62
Muir, John, xvi, 35, 49, 56, 60,
    80, 81, 201
Mumford, Lewis, 169
Murray, Bruce, 10, 12, 34–35
music of place, 164–68
My Lai massacre, 170
mysterium tremendum, 203

myth(s), 21; of America, 42–43,
    51, 52, 134; of American West,
    xii, 71–73; of Great Outdoors,
    1–2

Nantucket Island, 50–51
Nashville, 47, 88
National Mall, 47
national parks, xvi, 47, 58,
    217n69
Native Americans, 78
Natural Resources Conservation
    Service, 188
nature, natural world, xv–xvi,
    xvii, 32, 56, 58, 107, 128, 129,
    197, 201; accepting, 190–91;
    American identity and, 53–54;
    American perceptions of,
    20–21, 23–24, 82; bristlecone
    as symbol of, 180–85; conquest
    of, 104–5; engineering and,
    108–9, 111; vs. industrializa-
    tion, 48–49; interpretation
    of, 59, 61; inventions and,
    113–14; in landscape art,
    52–53; and magic lantern
    shows, 62–64; Manifest
    Destiny and, 91–92; as Other,
    205–6; as physical barriers,
    119, 121; picturesque, 69–70;
    the sublime in, 66–69; and
    technology, 117–18; visibility
    of, 185–87
Nature Conservancy, 190
Nebraska, 51, 190
neighborhood mapping, 158–60

Neuromancer (Gibson), 6

New Deal, 115

New England, 43, 48, 154; tourism promotion in, 50–51

New Hampshire, 56

New Jersey, xv, *109*

New Orleans, 88

New World, 23, 42, 43

New York, 57, 153

New York City, 96, 108, *109*, 169, 170; world's fairs in, 63, 88, 99–100

New York Times: on American travel, 46–47; on triumphalism, 143–44

Niagara Falls, 54, 55; engineering view of, 108–9

Nine Nations of North America (Garreau), 51–52

Niobrara River, 190

Northern Pacific Railroad, 71

Novak, Barbara, 54

Novak, Marcos, 19

"Nowhere Man" (Beatles), 177

O'Keeffe, Georgia, 75

Old North Church, 47

Olmstead, Frederick Law, 49, 60, 70

Omaha, world's fair in, 88, 91–92

Oregon, 137

O'Sullivan, John, 136, 137

Other, nature as, 205–6

Otto, Rudolph, 32, 203–4

outdoors, American perception of, 1–2

The Oxbow (Cole), 52, 53 (fig.)

Padua, Arena Chapel in, 22

Palmer, Frances, 109

Pan-American Exposition, 90

Panorama of the Mississippi River (Banvard), 62

panoramas, 59; mechanical, 61–62

Parker, Horatio, 167

Parkes, Henry Bamford, 127–28

Park Forest, 173

parks, 70

parlors, 1–2

Pascal, Blaise, 150

Patriarch Grove, 179

patriotism, 46, 53

"Paul's Case" (Cather), 105

Peirce, Charles S., 160

Pennsylvania, and September 11 attacks, 170

Pennsylvania turnpike, 121

Pentagon, September 11 attacks on, 170

People of Plenty (Potter), 113–14

permanence, 199; vs. placelessness, 197–98

Philadelphia, Rudolph Hering in, 110–11

Philadelphia Centennial Exposition, 48, *85*, 86–87, 88, 92

photography, 4, 5, 39; and journeys, 40–41; and magic lantern shows, 62–64; stereoscopic, 64–66

physiographic maps, 187–88

picturesque in art, 69–70

Pike, Zebulon, 75

Pike's Peak, 58

pilgrimages, 54; to the West,
71–73
Pilgrim at Tinker Creek (Dillard),
201–2
Pinchot, Gifford, 115
Pittsburgh, 48, 122
place, xiii, xviii, 149–50, 176,
179, 197, 198; in American
history, 192–95; defining,
151–53; evil, 168–70; formative experience of, 200–201;
and freedom of movement,
171–72; importance of, 31–32;
inhabitation with, 199–200;
interchangeable, 174–75; music of, 165–68; orientation by,
154–55; role of, 202–3; sacred,
203–5; stories of, 162–64; in
Third Nature, 155–57
placelessness, 171, 176; as global
village, 177–78; vs. permanence, 197–98
play, playfulness, 16–17, 29
Pollan, Michael, 204
pollution, 122, 168, 169
Polynesians, xiii
Populus, 96
Portland OR, world's fair in, 88,
91, 92
Potter, David M., *People of Plenty*,
113–14
poverty, 126
Powell, John Wesley, 77, 78, 79
Presley, Elvis, 47
private property: ecology and,
196–97; land ethic and,

193–94; legislation, 197–98;
value of, 195–96
Progressive Era, 115
Project Entropia, 17
property law, 197–98
Prospect Park, 110
Protestantism, 128
pseudo-events, 59
Puritans, 154
Purple Rose Syndrome, 6, 17
Putnam's Monthly Magazine, 57
Pyne, Steve, *How the Canyon
Became Grand*, 78

Radio City Music Hall, 23
Radosh, Daniel, 96
Raid on Bungeling Bay, 15
railroad networks, 119, *119*
railroads, 57(fig.), 59, 94, 112,
116(fig.), 134; impact of,
117–19; and national parks,
58, 71
Raisz, Erwin, physiographic maps
by, 55, 187–88
Rashomon Effect, 30–31
reality, 9–10, 102; and memory,
160–61; normal and nonnor-
mal, 28–29; and virtual reality,
33–34
recreation, xvi, 81
re-creations, 46
Redeemer Nation, 138
redwoods, 47
Renaissance, 42, 43, 87–88
Rensselaer, Mariana G. van, 91
resorts, 56, 58(fig.), 71–72, 80

Rettberg, Scott, 20
Ricoeur, Paul, 204–5
Roebling, John August, 108, 109–10
Rollins, Frank, 50
Rome, as theme of Columbian Exposition, 92–93
Roosevelt, Franklin D., 115, 141–42
Roosevelt, Theodore, 80, 115, 134, 141
Root, Elihu, 141
"Rosebush Exercise," 25–26
Roth, Roxy, 22–23
Rousseau, Jean Jacques, 125
Runte, Alfred, 71
rural life, 126; landscapes of, 2 (fig.), 50–52, 132; tourism and, 49–50
Rushdie, Salman, 174
Rydell, Robert W., 83, 87, 98

Sacks, Oliver, 27, 31, 35, 37; on reality, 28–29
sacred in space and time, 32
sacred places, 54, 169, 203–5; deserts as, 76–79
Sade, Marquis de, *Histoire de Juliette*, 22
Saint Louis, world's fair in, 88, 92
Saint Mark's (Venice), 22
San Antonio, 88
Sand County Almanac (Leopold), 193
Sanders, Scott Russell, xvi–xvii, 172, 191, 206

Sandin, Dan, 19, 20
San Francisco, world's fairs in, 88, 91, 92
Satt, Hilda, 94
scenic beauty, importance of, 194–95
Schiesel, Seth, 14
Schor, Juliet, 130
Schulman, Edmund, on bristle-cone pines, 178–79
Schulmeyer, Gerhard, 146
Scientific Revolution, 125
Seattle, 88
Second Nature, xv, 7, 95, 107, 111, 113, 130–31, 137; infra-structure and, 114–15, 123
Sedgwick, Catharine, 45–46
Sedgwick, James, 63–64
seizures, nature of, 35
sense of place, ix, xiii, 107–8, 201
September 11, 2001, attacks, 170
Shaub, George, 41
Shepard, Paul, on placelessness, 176–77
Rock with Wings (Shiprock), 74
shopping malls, 102–3, 104, 174
shrines, national historic, 47
Sierra Club, 80
Sierra Nevada, Bierstadt's paint-ings of, 65
Silko, Leslie Marmon, 162
SimCity, 15, 95, 96
Simon, John, 20
The Sims, 13–14, 97
simulations, 17–18
The Sirens of Titan (Vonnegut), 33

Skiff, F. J. V., 92
skyscrapers, 94, 113
Slossen, Edwin P., 104
slums, 122
Smith, Jedediah, 72
Smith, John, 43
Smoky Mountains, 47–48
Sobel, David, 157
Social Darwinism, 129
social mobility, consumerism and, 126, 127
social structure, Internet and, 10, 12
Soil and Water Conservation Service, 188
"Song of Myself" (Whitman), 25
Soviet Union, 99, 142
space, xii, xiii, 45, 131
Space and Place (Tuan), 155–56
Specimen Days (Whitman), 24–25
spirit-flesh dualism, 43
spirituality ecoregions, 190
Spokane, 88
steamboats, 71, 116, 117
steam engine, Corliss, 85, 86
steel works, 48
Stein, Gertrude, 174
Steinberg, Ted, 197
stereopticon, 1–2
stereoscope, 63, 64–66
stereotypes, landscape as, 29–30
Stilgoe, John R., xvii–xviii, 29, 200
Stockton, Richard, 20
stories, place in, 162–64
Strong, Josiah, 140

sublime, 200; in landscape art, 66–69
suburbs, mobility in, 173–74
Superfund sites, 169
superpower, U.S. as sole, 142
Symphony No. 3 (Ives), 167

Tall, Deborah, xvi, 174, 175, 195, 205
technology, 134, 138; consumerism and, 124–25; nature and, 117–18; in video games, 14–15
telautograph, 93
telegraph, 134
telephones, 94
television, 4
Tennessee Valley Authority (TVA), 115
Terry, Sara, 131, 132
Tetsuro, Watsuji, 199
Texas, 137
textile mills, 48
theme parks, 46
Theory of the Leisure Class (Veblen), 129
"The Things Our Fathers Loved" (Ives), 165–66
Third Nature, 7, 34, 107; place in, 155–57
Thoreau, Henry David, 23, 24, 25, 49, 54, 59, 192; Walden, 127
Three Places in New England (Ives), 166
time, in desert, 76, 78
Times Square, 47

Tinker Creek, xvi, 32, 175, 201–2
Tocqueville, Alexis de, 29; on American exceptionalism, 138–39; *Democracy in America*, 126–27
topophilia, 203
totalitarianism, 99
tourism, 103; American, 45–49, 55; eco-, 80–81; European, 43–44; Grand Tour, 56–58; nature and, 58–59; rural landscape, 49–52; the West, 70–73
Townsend, George Alfred, 87
traditionalism, Western, 73
Transcendentalism, 154; nature and landscape in, 54, 56
transportation networks, American, 117–18
travel: in America, 46–49; virtual, 17–18
travel talks, 63
Treblinka, 169
trees: aesthetics of, 182–83; bristlecone pine, 178–85; contemplation of, 201–2; sacred, 54
Trenin, Dimitri, 142
Triumphal America, xii–xiii, 108, 134, 140–41
triumphalism, 134, 137–38, 140–41, 142; twenty-first century, 143–45
Trollope, Frances, 44
True Crime: New York City, 96
Tuan, Yi-Fu, xii, xiii, xiv, 72–73, 154, 157, 172, 176, 203; on inhabitation of space, 199–200; *Space and Place*, 155–56

Tucson, 76
Turkle, Sherry, 6, 17, 18, 34, 38
Turner, Frederick Jackson, 73, 128, 140, 153, 162, 172, 192; on Mark Twain, 163, 164
Tuveson, Ernest Lee, 139
TVA. *See* Tennessee Valley Authority (TVA)
Twain, Mark, 51; on consumerism, 127–28; on Hannibal, 162–64
2001: A Space Odyssey (Kubrick), 33
Tyler, John, 137
Tyndall, Mount, 69
typhoid epidemics, 122

Ultima Online, 18
Union Pacific Railroad, 64, 118
Union Stock Yards, 48
Universal Studios, 46
UOTreasures, 17
urban sector, 115, *120–21*, 169; interchangeable landscapes of, 174–75
USA Today, 143
U.S. Department of Agriculture, 188
U.S. Geological Survey, 188
Utah, *74, 75*
utopias, utopianism, 99, 104, 105, 153–54; metropolitan, 95–96

Van de Water, Frederic, 72
Variations on "America" (Ives), 165
Vaux, Calvert, 70

Veblen, Thorstein, 129–30
Venice, Saint Mark's in, 22
video games, 12, 17; childhood
    and, 13–16; virtual reality of,
    18–19
View-Masters, 63, 66
VirtuaLand, 18
virtual reality, ix, x, xii, xiii–xiv,
    7, 9, 17, 22, 32, 36, 46, 107,
    150; America as, 34–35, 44;
    cognitive mapping as, 26–28;
    development of, 4–6; fairs as,
    87–93; Grand Tour as, 56–58;
    of history, 20–21; imperma-
    nence of, 37–38; money in,
    145–46; nature as, 58–59;
    place in, 155–56; reality in,
    33–34; video games, 18–19
Vonnegut, Kurt, 33

Walden (Thoreau), 127
Walden Pond, 54, 59
Walrus, 9
war, warfare, 138; apocalyptic,
    139–40
Warner, Sam Bass, 123
War of 1812, 170
war simulations, 36
Wasteland (Eliot), 177
water systems, 122
Watt, James, 84
Wattles, Gurdon, 91
Wayland, Francis, Elements of
    Political Economy, 127
Weber, Max, 204
Web sites, 11, 42, 157
Wells, David A., 104

Wells, H. G., The World Set Free, 105
Welty, Eudora, 162
Wentworth, John, 137
Wertheim, Margaret, 21–22
West: land surveys of, 79–80; as
    Original America, 70–71; tour-
    ism in, 70–73
West Edmonton Mall, 102–3
western expansion, 21, 62,
    172–73
Western Sea, 43
Westinghouse, George, 84, 93–94
whaling industry, as romantic,
    50–51
Wheeler, George Montague, 79
Wheeler Peak, 181
White, Lynn, Jr., 112
White City, 84, 88, 89, 90, 94
White Mountain CA, 179
White Mountains, 56
White Sulphur Springs, 56
Whitman, Walt, 23; "Song of My-
    self," 25; Specimen Days, 24–25
Whittredge, Wyrthington, 44
WildEarth movement, 80
wilderness, 52, 54, 60, 133, 194;
    Grand Tour and, 56–57
wilderness appreciation, 81
Wilkins, James F., Moving Mirror of
    the Overland Trail, 62
Williams, Terry Tempest, 184–85
Wilson, Eric, 24, 25
Wilson, Woodrow, 134, 138, 141
Wisconsin Supreme Court, 196
Wolcott, Jennifer, on eating
    locally, 131–32, 134
Wolf, Joshua, 102

Wolfe, Thomas, 159
The Wooing of the Earth (Dubos), 198
World Columbian Exposition, 87, 88, *89*, 90, 91, 140; electric lighting at, *84*, 93–94; Rome as theme of, 92–93
World of Tomorrow, 99–100
The World Set Free (Wells), 105
world's fairs, 48, 63, 104; American, *85*, 86–95, 98–100; London, 83–84
World War I, 138, 141
World War II, 99–100, 142, 154, 155
World Wide Web, 8, 90
worm gears, 36

Worster, Donald, 77
Wounded Knee, 72, 170
Wright, John K., 203
Wright, Will, 12, 15
Wrigley Field, 47
Wyeth, Andrew, 76

Yellowstone, 47, 54, 55, 58, 61, 72, 110; greater ecosystem of, 81, 188–90
Yosemite, xvi, 47, 54, 55, 58, *60*, 61, 65, 68, *68*, 72, 80
Yosemite Valley, California (Anthony and Anthony), 65

Zeilinger, Anton, 9–10
Zion National Park, 21, 32, 81